Surveys and Tutorials in the Applied Mathematical Sciences

Volume 9

Featuring short books of approximately 80-200pp, Surveys and Tutorials in the Applied Mathematical Sciences (STAMS) focuses on emerging topics, with an emphasis on emerging mathematical and computational techniques that are proving relevant in the physical, biological sciences and social sciences. STAMS also includes expository texts describing innovative applications or recent developments in more classical mathematical and computational methods.

This series is aimed at graduate students and researchers across the mathematical sciences. Contributions are intended to be accessible to a broad audience, featuring clear exposition, a lively tutorial style, and pointers to the literature for further study. In some cases a volume can serve as a preliminary version of a fuller and more comprehensive book.

Marek Stastna

Internal Waves in the Ocean

Theory and Practice

Marek Stastna
Department of Applied Mathematics
University of Waterloo
Waterloo, ON, Canada

ISSN 2199-4765 ISSN 2199-4773 (electronic)
Surveys and Tutorials in the Applied Mathematical Sciences
ISBN 978-3-030-99209-5 ISBN 978-3-030-99210-1 (eBook)
https://doi.org/10.1007/978-3-030-99210-1

Mathematics Subject Classification: 76B25, 76B55, 76F45

This Springer imprint is published by the registered company Springer Nature Switzerland AG
The registered company address is: Gewerbestrasse 11, 6330 Cham, Switzerland

For my sister Kazi, who has always been the best writer in the family.

Preface

The study of the ocean has changed in a fundamental way over the past 50 years. The technological revolution ushered in by the end of the Second World War has revolutionized how we measure and conceptualize the oceans. During my graduate work in the late 1990s, it was still common to have entire quantitative courses devoted to perturbation theoretic approaches, as exemplified by Pedlosky's classical treatise *Geophysical Fluid Dynamics*. The very regular oceanic circulations that typify the results of such theories were already known to be largely wishful thinking of the human mind, though it was the worldwide dissemination of the results of the TOPEX/Poseidon satellite that confirmed the notion that the ocean was a field of eddies. In the subsequent decades, the scales of simulation and observations have become even finer, with eddy permitting simulations yielding to true eddy resolving simulations, and the revealed mysteries of the mesoscale supplanted by the emerging complexities of the submesoscale. I believe variants of this story are widely reported and the sentiments widely believed—or as is more typical of academics, not disagreed with too strongly. It is worth pointing out that even without computation, it is possible to present the motion of fluids in an accessible way that is rooted in reality [1].

In fact, there is an older story, on even smaller scales that better exemplify the co-mingled evolution of science, computation, and applied mathematics in the late twentieth century: this is the story of internal solitary waves. The story of solitary waves on the surface of water is relatively well known, with John Scott Russel's horse ride along a Scottish canal a part of the popular legends of applied mathematics:

https://en.wikipedia.org/wiki/John_Scott_Russell#The_wave_of_translation

The scientific world of the mid-nineteenth century was not ready for a nonlinear wave of permanent form, and Russel did not live to see the acceptance of his discovery by the academic world. Indeed, even the 1895 discovery of the equation governing Russel's wave of translation barely made a ripple in the scientific consciousness of the time, something that is hard to believe given today's ubiquity of the eponymously named Korteweg–de Vries (KdV) equation. Johnson's book [2] on nonlinear theories of water waves is concerned with surface, as opposed to internal, waves, but has excellent historical notes relevant to both.

The KdV equation returned to the scientific consciousness as part of the Fermi-Pasta-Ulam-Tsingou problem, one of the first examples of computational physics, but even this was a "limited audience" return. Things changed rather fundamentally in the late 1960s as part of the discovery of a solution procedure that allowed for an analytical solution of the nonlinear KdV equation. The procedure, called the inverse scattering formalism, is an astonishing tour de force of applied mathematics. For many people, myself included, it is one of the highlights of twentieth-century applied mathematics. During the 1970s, with extra motivation provided by Cold War rivalries, this method spawned an incredible array of ingenious generalizations and extensions. Over the course of my career, it has been a true pleasure to occasionally interact with individuals (sometimes in person, sometimes through peer review) who were part of this mathematical revolution. And yet, when I am honest with myself, today's world of ubiquitous computing could well ignore all this mathematical beauty and continue on its way unhindered. No one, in my opinion, presents this exploration-centered point of view on mathematics than Nick Trefethen, and his beautiful little book [3] provides a gateway to the many avenues of exploration discussed in this paragraph.

Were this the end of the story, there would be no need to read much further. Two developments changed the game, as far as internal solitary waves were concerned. The first was David Benney's derivation of the KdV equation from the full stratified Euler equations using perturbation theory in wave amplitude (nonlinearity) and the aspect ratio (dispersion) [4]. The second was the development of fast Fourier transform-based (FFT) spectral methods for the solution of the KdV equation, for example, by Fornberg and Whitham. Both of these are readily recognizable in the literature of 2019, and indeed in the material in this book. They are examples of how choice of "notation," in mathematics and in algorithms, has a profound influence on uptake. There is a long history of such issues, dating back to the discovery of calculus and the superiority of Leibniz's notation over Newton's. Indeed, one can analogize today's movement to open source software through GitHub and similar platforms to this as well.

The rapid developments in applied mathematics and computation from the 1960s onward were mirrored by improvements in other disciplines, and ocean observation and measurements were no exception. At times motivated by commercial applications (e.g., oil exploration and extraction), the in situ measurement of internal solitary waves picked up pace through the 1980s and 1990s. This was aided and abetted by space-borne observations: the typical light and dark bands observed near straits (e.g., Gibraltar, Messina, and Luzon) that are the surface signatures of the otherwise invisible-to-the-eye waves in the ocean interior. Observations are difficult to publish without some way to draw in the readers' (really the reviewers') interest, and hence large-amplitude internal waves, often labelled as internal solitary waves, have been linked with enhanced nutrients, irreversible mixing, and transport across the boundary layer. Quite a trip for a phenomenon known to the ancient Norse, but believed to involve magical fish! The online version of Alex Korobov's PhD thesis has excellent historical notes (including the magical fish):

```
https://uwspace.uwaterloo.ca/handle/10012/3170?show=full
```

This book is about the mathematical description of internal solitary waves. It is however most certainly not a mathematics book. Modern mathematics can be extremely exclusionary, with notation and vocabulary that long ago lost their link to the physical world. Moreover, the same Gaussian theorem–proof tradition that is the basis of modern mathematics forces the presentation style to follow a logical path that is the opposite of the inductive route along which scientific discoveries are made. At the same time, the raw power of modern computing makes it possible to confirm mathematical statements relevant to the scientist in a very short amount of time. Indeed there are calculations I present for the reader in this book that were carried out overnight when I was working on my PhD thesis, and yet now their result is ready for visual presentation on a simple laptop in minutes. Thus, the question is no longer "can we" but "should we."

Many of the theories of internal solitary waves are impressive feats of applied mathematics, but precisely because they are linked with powerful mathematics, their physical relevance is never tested. On multiple occasions, I have been asked by a reviewer whether my exact calculation agrees with the reviewer's preferred approximate theory, and yet to this day the vast majority of theoretical papers on internal solitary waves contain no validation against data. At the same time, it must be said that the mathematics provides an essential order beyond just cataloguing the interesting observations of internal solitary waves, and thus I have a lot of sympathy for the colleague obsessing over an arcane mathematical observation like a nanopteron or a breather even though observations are unlikely to yield one outside of exceptional situations. I hope to convey some of this balance to the reader.

Thus, this is a bit of a subversive book. I would like the reader to try the codes; change the codes and at some point supercede the codes. I also hope they will find the style of the codes somewhat unique. I am heavily influenced by Nick Trefethen's approach to code (flat and succinct, though perhaps with a few more comments than some of his Matlab gems). There is a role for proper software carpentry, the use of libraries, and the most modern HPC architectures, but so much can be done locally, with Matlab or one of its free alternatives, so I mostly stick to that in what follows. I would also like the reader to think about scales. Even though the objects of study in this book are large compared to us (tens to hundreds of meters in typical length), they are tiny on the scale of ocean models, and indeed the vast majority of ocean models cannot represent them at all. Thus, I ask the reader to take up the struggle of imagining a large, complex world from a point of view that differs from any daily experience. It is my belief that this struggle is at the core of the rational person's approach to modern environmental science. Perhaps every prospective oceanographer should don a VR headset and get virtually dropped off in the central Pacific with the ability to move at the speed of highway traffic, so that they can only proceed to "real" science when they've gained an appreciation for the incredible size of the ocean. Of course, I have no means to do this, so instead I ask the reader to develop their intuition via the attached codes.

I am indebted to Kevin Lamb (Waterloo) for teaching me much of this material first as a PhD advisor, and later as a colleague. The notation for weakly nonlinear theory follows his presentation almost exactly (all mistakes are my own). Magda Carr has enriched my point of view with the point of view of a first-rate experimentalist, and the footprints of our collaborations and her own papers can be found throughout this book. I appreciate the gentle prods from Peter Diamessis (Cornell), Justin Shaw (Waterloo), and Donna Chernyk (Springer) that got me to start writing this material. My group at Waterloo, the aforementioned Kevin Lamb, Francis Poulin, and Mike Waite provide an outstanding intellectual environment with plenty of mutual support, and the students that have come through the group contributed much of the raw material for this work through their questions, and in some cases their own codes. Particular thanks go to Chris Subich, Derek Steinmoeller, Michael Dunphy, and Kris Rowe for their philosophies and opinions on code; Nancy Soontiens, David Deepwell, Andrew Grace, and William Xu for their ideas on internal waves; Aaron Coutino and Jared Penney for ideas about heat–salt systems; and Justin Shaw for ideas about data. Nico-Castro Folker selflessly volunteered to act as a guinea pig/informal editor for the classroom application of the book. Finally, my wife Myra Fernandes, and her own set of academic interests, has provided me with decades of new ideas, and beyond academics, the support needed to live a real life.

Reader's Guide

This is not a long book, and I hope its style makes it relatively easy to read. Nevertheless, modern life is busy, and the modern student is not always "linear" in their approach to learning. In fact, I find I rarely read technical subjects cover to cover, following the precise order of presentation. In Fig. 1, I present what I consider the essential aspects of this work, and some of the choices a reader can make. In particular, I would like readers to focus on the active parts of the book, where codes to explore concepts are provided and learning is done by examining the figures these codes produce. This is not to denigrate the theory behind it all, but algebra tires the mind, and two decades of teaching experience has shown me that the vast majority of students reach the end of a complex calculation and are too tired from the effort to probe the physics of what the calculation reveals. For this reason, the theoretical chapter is chosen to provide well-chosen examples of what students may be missing, for which all the details are provided, but to only hint at the many, more complicated extensions available in the literature.

The two extension chapters are meant to make some contact with the literature and disabuse the reader of the notion that "it's all been done." The world's oceans are vast, and our measuring devices (while ingenious) sample only very sparse regions

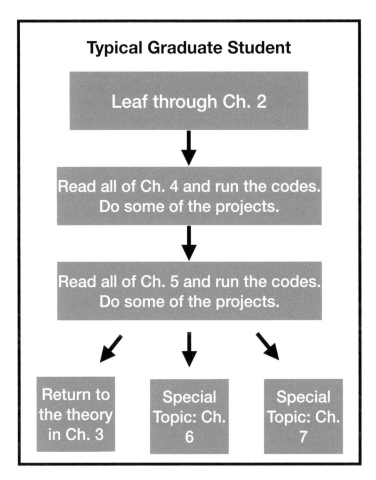

Fig. 1 A schematic of a typical reader's route through the book

of them. Moreover, in many ways, we lack the mathematical language to describe what is really going on in such a complex, nonlinear environment.

Please download the MATLAB codes from GitHub:

https://github.com/mmstastn/internal_waves_in_ocean_book

Literature Preface

While I do not wish to conduct an authoritative literature review, I do wish to provide some references for further reading and some background. I will do so on a chapter by chapter basis, but will number consecutively. For this first chapter, I wish to identify three books and one research paper that have influenced the thinking behind the construction of this book.

1. Kundu, P.K., (1990) *Fluid Mechanics*, Academic Press, San Diego.

   No science related book has had a larger influence on my conceptualization of fluid mechanics than the first edition of this book. After P.K. Kundu's untimely passing, the book has undergone many changes through multiple editions that have largely left it unrecognizable. The first edition reflects the author's passion for fluid mechanics seen through the prism of the natural environment, and I don't think I go a week without leafing through its pages. The end of chapter notes are often a treasure trove of older papers well worth reading.

2. Johnson, R.S. (1997) *A Modern Introduction to the Mathematical Theory of Water Waves*, Cambridge Press, Cambridge, UK.

   A tour de force of theory, with the poetic version of the Legend of Arthur heading each chapter, this book is very much something I could not write. But, with its many exercises, it did allow me to develop as a student of wave theory and forge my own identity as a numerical experimentalist. The Bibliography is impressive in coverage (e.g., Russel's original report on the observation of a solitary wave, the original Korteweg and de Vries paper, and the original Fermi, Pasta, Ulam report), and the historical notes cover many of the players in classical wave theory.

3. Trefethen, L.N. (2000) *Spectral Methods in MATLAB*, SIAM Press, Philadelphia.

   A close second, in terms of personal influence, is this tidy little gem by Nick Trefethen. The idea that real science and real mathematics can fit on a (small) single page of code was so provocative that when I first found the book, I used to read it "in hiding," as if the professors that taught me functional programming would find me and cast a disapproving eye! The references, while not as tidy in terms of presentation, are excellent (e.g., the Fornberg and Whitham paper mentioned above).

4. Benney, D. J. (1966) "Long Nonlinear Waves in Fluid Flows," *J. Math. Phys.*, 45, 52–69.

   The notation is amazingly fresh for a paper that is now more than 50 years old.

Waterloo, ON, Canada   Marek Stastna

Contents

Chapter 1
Background and Equation Summaries

In this chapter, we provide a succinct exposition of the background material. Most readers will have their own collection of favorite textbooks, and thus I provide a single reference [5], for the background mathematical material. [1] and [2] provide more than enough on fluid mechanics for any reader. A unique aspect of this chapter is a compendium of the relevant sets of equations relevant to the remainder of the book. The sets of equations are presented in color-coded boxes with sets of governing equations presented in gray scale, approximate solutions presented in warm colors, and solutions presented in cool colors. In later sections, mini-projects are put into soft yellow boxes.

Mathematics Background

While the majority of the portions of this book assumes only what is commonly referred to as "mathematical maturity," it is worth reviewing some basics.

It may seem like a trite statement, but to consider solutions of a differential equation, we must ensure that the quantity in question may be differentiated enough times to make the differential equation meaningful. This is a significant concern in nonlinear hyperbolic systems, such as the equations of gas dynamics or shallow water equations, which may form shocks in finite time. However, it will not be a concern for us, since in the material discussed in this book nonlinearity may be balanced by dispersion.

Second, recall from vector calculus that a vector field $\vec{u}$ is **incompressible** provided $\nabla \cdot \vec{u} = 0$ and **irrotational** provided $\nabla \times \vec{u} = 0$. In two dimensions (for example, the frequently used $x - z$ plane), which we will consider for the majority of the examples in this book, an irrotational velocity field implies that $\vec{u} = \nabla \phi$, or in other words that the velocity field is derived from a potential ϕ. Similarly, an incompressible velocity field implies the existence of a "streamfunction," ψ, so that

M. Stastna, *Internal Waves in the Ocean*, Surveys and Tutorials in the Applied Mathematical Sciences 9, https://doi.org/10.1007/978-3-030-99210-1_1

the velocity field is parallel to lines of ψ =constant, or in other words

$$\vec{u} = (u, w) = \left(\frac{\partial \psi}{\partial z}, -\frac{\partial \psi}{\partial x} \right).$$

A further calculus fact we will use is that the fact that for functions of two variables, $f(x, z)$ and $g(x, z)$, if the Jacobian

$$J(f, g) = \frac{\partial f}{\partial x} \frac{\partial g}{\partial z} - \frac{\partial f}{\partial z} \frac{\partial g}{\partial x}$$

vanishes, then the two functions are dependent,

$$f = G(g)$$

for some (unknown) function G.

While not necessary for the majority of the material, the DuBois-Reymond Lemma of Calculus of Variations sits in the background of all Continuum Mechanics theories. It has many different formulations, but the one relevant to Continuum Mechanics states that if

$$\int \int \int_D f(\vec{x}) dV = 0$$

for all possible D, then $f(\vec{x}) = 0$ pointwise. Technical details involve the properties D must have, though for Continuum Mechanics D smooth, and simply connected is a safe assumption.

Finally, the theory of Sturm–Liouville problems turns up in a somewhat non-standard way. Sturm-Louiville theory is a common topic in many books on partial differential equations (see Strauss for a nice introduction). A second order ODE

$$\frac{d}{dx} \left(p(x) \frac{du}{dx} \right) + q(x)u + \lambda \sigma(x)u = 0 \tag{1.1}$$

with the homogeneous boundary conditions

$$\alpha_i u(a_i) + \beta_i u'(a_i) = 0 \text{ where } i = 1, 2 \tag{1.2}$$

is said to be of Sturm–Liouville type if $p(x)$, $p'(x)$, $q(x)$ and $\sigma(x)$ are continuous with $p, \sigma > 0$ on $[a_1, a_2]$. The problem is often written as

$$\mathcal{L}u = -\lambda \sigma(x)u.$$

The problems we will see will have Dirichlet boundary conditions so that $\beta_1 = \beta_2 = 0$. Sturm–Liouville problems are a generalization of Fourier series, since it can be proven that the set of eigenfunctions is complete and orthogonal, with an inner product induced by the form of the particular Sturm–Liouville problem in question. This in turn allows for the definition of an adjoint operator $\mathcal{L}^\dagger$, so that self-adjoint problems for which $\mathcal{L} = \mathcal{L}^\dagger$ may be defined. A standard example of a self-adjoint operator is

$$\mathcal{L} = \frac{d^2}{dx^2}$$

for the standard inner product

$$\langle f, g \rangle = \int_a^b f(x)g(x)dx$$

Now, the *Fredholm Alternative* is likely familiar to the reader from Linear algebra where it states that if $\mathcal{A}\vec{x} = \vec{0}$ has only the trivial solution, then $\mathcal{A}\vec{x} = \vec{b}$ has a unique solution, as well as the converse (i.e., it is an *iff* statement). A far less common version states that $\mathcal{A}\vec{x} = \vec{b}$ has a unique solution *iff* $\langle \vec{b}, \vec{v} \rangle = 0$ for all $\vec{v}$ that satisfy $\mathcal{A}^*\vec{v} = 0$ where $*$ denotes the matrix adjoint.

The Sturm–Liouville version states that the boundary value problem $\mathcal{L}u = f$, $\alpha_i u(a_i) + \beta_i u'(a_i) = 0$ where $i = 1, 2$ has a unique solution *iff* $\langle f, v \rangle = 0$ for any solution of the problem $\mathcal{L}v = 0$.

Wave Theory Background

Wave phenomena are a broad and important part of fluid mechanics and physics in general. The basic definitions of waves are usually covered in first year physics courses. As a brief review, if we consider a propagating, sinusoidal wave in one spatial dimension with the form $\cos(kx - \sigma t)$ (sometimes called a plane wave) where x is the space variable and t is the time variable, then k is called the **wave number** and σ is called the **frequency**. The **wavelength** is defined as $\lambda = 2\pi/k$ (typically measured in meters), and the **period** of oscillation is defined as $T = 2\pi/\sigma$ (typically measured in seconds).

In mathematics (and again in one spatial dimension for simplicity), waves are often associated with solutions of the **wave equation** (let subscripts denote partial derivatives),

$$A_{tt} = c^2 A_{xx}$$

or its uni-directional cousin

$$A_t = -cA_x.$$

Both of these equations are formally referred to as hyperbolic equations, though may be a more accurate description would be "signaling equations" since a signal is carried along special directions called the *characteristics*.

However, plane waves, like the one given in the first paragraph of this section, are a more general part of physics and can often be the solution to partial differential equations (PDEs) other than the wave equation.

Let us consider a toy problem,

$$A_t = -cA_x + \alpha A_{xxt}$$

where $\alpha > 0$ is a physical parameter. If we assume a solution of the form

$$A = a_0 \exp[i(kx - \sigma t)]$$

the cosine, or plane wave, solution $\cos(kx - \sigma t)$ can be recovered by taking the real part. Thus the complex exponential is just a notation tool, saving us a bit of time with the algebra involved. The plane wave solution is sometimes called the **"wave ansatz"** in PDE books or the classical reference by Whitham [6]. Now, we need to make sure that the solution actually works. To this end, observe that the governing equation is linear so that after you take the various derivatives, every term will have a common multiple $\exp[i(kx - \sigma t)]$, and hence this term can be cancelled. That means that in order to determine if there is a possible wave solution, we must find a relation for the frequency as a function of wavenumber (or alternatively wavenumber as a function of frequency). A bit of algebra shows

$$-i\sigma = -cik + i\alpha k^2 \sigma$$

which upon rearranging gives

$$\sigma(k) = \frac{ck}{1 + \alpha k^2}.$$

The solution for $\sigma(k)$ is called the **dispersion relation**, and waves for which σ is not a multiple of k are called **dispersive waves**. Our example yields dispersive waves. For dispersive waves, the individual crests propagate with the phase speed $c_p = \sigma/k$, while energy (or a wave packet in physics parlance) propagates with the group speed $c_g = d\sigma/dk$. The two are generally not equal so that the individual crests in a wave packet either run ahead of the packet or disappear out of the back of the packet. For our example

$$c_p = \frac{c}{1 + \alpha k^2}$$

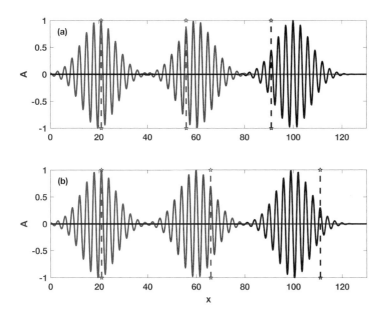

Fig. 1.1 (a) Wave packet propagating left to right at three different times with $c_p < c_g$ the position of a crest chosen at the initial time is indicated by vertical dashed lines. (b) as (a) but with $c_p > c_g$

so that crests propagate slower as k increases or wavelength decreases. A schematic of the evolution of a rightward propagating wave packet in which the position of a crest is indicated by a vertical dashed line is shown in Fig. 1.1.

Both the dispersive and hyperbolic equations discussed above are linear problems. In fact, as an initial value problem on the whole real line

$$A_t = -cA_x$$

with $A(x, 0) = F(x)$ has a famous solution due to D'Alembert $A(x, t) = F(x - ct)$. If we now consider a model nonlinear equation

$$A_t = -(c + \alpha A)A_x$$

we could ask if this equation has a special solution. The answer is "No," but we could use D'Alembert's solution to develop a bit of intuition on the effect nonlinearity has

$$A^{(trial)}(x, t) = F[x - (c + \alpha A)t].$$

The reader can see that this is in fact an implicit expression. However, taken at face value, it states that the propagation speed increases as A increases. Or in other words, the "tall" part of a wave travels faster. This is what leads to the steepening of surface waves on a good surfing beach and mathematically may be formalized using

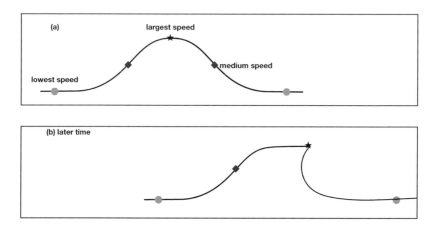

Fig. 1.2 The schematic of the nonlinear evolution of a wave with an amplitude dependent propagation speed. (**a**) An initial, symmetric disturbance with three heights/propagation speeds indicated. (**b**) The asymmetric disturbance at a later time. The wave is past breaking at this time

the theory of characteristics, where wave breaking corresponds to characteristics crossing. A schematic of the evolution of a nonlinear wave, as one might see on a surfing beach, is shown in Fig. 1.2.

The two phenomena of nonlinear steepening and dispersion are an essential aspect of the natural world, and indeed, it is the competition between the two that makes all the phenomena discussed in this book possible.

Fluid Mechanics Background

While it is impractical to include a full exposition of an introductory fluid mechanics course, we briefly summarize the background herein. From a continuum mechanics point of view, the fundamental notion is Euler's fluid particle, which is assumed to be small enough to be treated as a singleton, but large enough to carry macroscopic properties (e.g., temperature). If we then have a field quantity, again let us take temperature as an example, we have $T = T(x(t), y(t), z(t), t)$ so that the rate of change of temperature following fluid particle can be derived by the chain rule of multi-variable calculus and is commonly labelled as the material derivative

$$\frac{DT}{Dt} = \frac{\partial T}{\partial t} + \vec{u} \cdot \nabla T \qquad (1.3)$$

where $\vec{u}$ is the fluid velocity. The field equations of fluid mechanics are then derived using the concept of **material volume** or a simply connected collection of fluid particles identified by their initial position and labelled $W(t)$ for all $t > 0$. This concept is used to prove the Reynolds transport theorem, which begins with a quantity

$$F(t) = \int \int \int_{W(t)} f(\vec{x}, t) dV$$

and states that

$$\frac{dF}{dt} = \int \int \int_{W(t)} \left[\frac{Df}{Dt} + f \nabla \cdot \vec{u} \right] dV \qquad (1.4)$$

under fairly mild technical conditions. For statements of conservation, the DuBois-Reymond Lemma of real analysis is used to pass from an integral expression to a pointwise partial differential equation. This is perhaps easiest to see for the Conservation of Mass. Let

$$M(t) = \int \int \int_{W(t)} \rho(x, y, z, t) dV \qquad (1.5)$$

so that in the absence of sources and sinks we have

$$0 = \frac{dM}{dt} = \int \int \int_{W(t)} \left[\frac{D\rho}{Dt} + \rho \nabla \cdot \vec{u} \right] dV. \qquad (1.6)$$

In the last step, I used the Reynolds transport theorem. The DuBois-Reymond lemma gives us the relevant PDE

$$\frac{D\rho}{Dt} + \rho \nabla \cdot \vec{u} = 0. \qquad (1.7)$$

When density changes are small, as is the case for naturally occurring air and water under most circumstances (there is a fair amount of subtlety here!), we may rewrite

$$\frac{1}{\rho} \frac{D\rho}{Dt} + \nabla \cdot \vec{u} = 0$$

and argue that the first term is negligible. This leads to

$$\nabla \cdot \vec{u} = 0 \qquad (1.8)$$

a statement of Conservation of Volume, or alternatively a restriction that the velocity field is **incompressible**.

The statement Conservation of Linear Momentum is somewhat more complex since we now have to balance the rate of change with applied forces. However, if the body forces are represented in a general way as the gradient of a potential and τ

represents the stress tensor, we have

$$\frac{D\rho\vec{u}}{Dt} + \rho\vec{u}\nabla\cdot\vec{u} = \nabla\cdot\boldsymbol{\tau} - \nabla\pi. \tag{1.9}$$

For Newtonian fluids, it is common to write $\boldsymbol{\tau} = -p\mathbf{I} + \boldsymbol{\sigma}$ where p denotes the pressure, $\mathbf{I}$ is the identity matrix, and $\boldsymbol{\sigma}$ denotes the viscous portion of the stress tensor. Moreover, the left hand side can be simplified using the Conservation of Mass to give a simplified set of equations

$$\rho\frac{D\vec{u}}{Dt} = -\nabla p + \nabla\cdot\boldsymbol{\sigma} - \nabla\pi.$$

If gravity is the only relevant body force, we have that $-\nabla\pi = -\rho g\hat{k}$ where $\hat{k}$ is the unit vector in the vertical direction directed upward and g is the acceleration due to gravity. Away from solid boundaries, the fluid viscosity for seawater, freshwater, and air is small, and hence, $\nabla\cdot\boldsymbol{\sigma}$ may be neglected. This leaves the stratified Euler equations for an inviscid fluid

$$\rho\frac{D\vec{u}}{Dt} = -\nabla p - \rho g\hat{k}.$$

The density of naturally occurring water, whether seawater or freshwater, is typically nearly constant ($\rho_0 = 1000\,\text{kg m}^3$ for pure water and $\rho_0 = 1023.6\,\text{kg m}^3$ for sea water). This allows for a simplification of the above equations, under the name "the Boussinesq approximation." The Boussinesq approximation posits that the variations of density are irrelevant except in the buoyancy term, leaving us the final momentum equation

$$\rho_0\frac{D\vec{u}}{Dt} = -\nabla p - \rho g\hat{k}. \tag{1.10}$$

These three equations are augmented by the Conservation of Mass

$$\nabla\cdot\vec{u} = 0$$

but this is not enough to uniquely specify the five variables $\vec{u}, p, \rho$. The final equation, often called the density equation, looks very similar to the Conservation of Mass

$$\frac{D\rho}{Dt} = 0 \tag{1.11}$$

but is in fact derived from energy considerations (textbook references are difficult to find, though see Mueller).

The most important equation derived from the momentum equations is that for vorticity. Vorticity is defined as

$$\vec{\omega} = \nabla \times \vec{u} \tag{1.12}$$

and in the $x - z$ plane $\vec{\omega} = (0, \omega, 0)$ where

$$\omega = \frac{\partial u}{\partial z} - \frac{\partial w}{\partial x}. \tag{1.13}$$

In three dimensions, the vorticity equation reads

$$\frac{D\vec{\omega}}{Dt} = \vec{\omega} \cdot \nabla\vec{u} + \frac{g}{\rho_0}\left(-\frac{\partial \rho}{\partial y}, \frac{\partial \rho}{\partial x}\right). \tag{1.14}$$

The left hand side represents the rate of change of vorticity following a fluid particle. The first term on the right hand side represents the well known vortex stretching/tilting mechanism that is fundamental for the understanding of turbulence, while the second represents the generation of vorticity due to density changes. The latter is often referred to as the baroclinic vorticity generation term [1]. In two dimensions, the equation simplifies considerably

$$\frac{D\omega}{Dt} = \frac{g}{\rho_0}\left(-\frac{\partial \rho}{\partial y}, \frac{\partial \rho}{\partial x}\right), \tag{1.15}$$

and we can see that ONLY variations in density can generate vorticity.

Stratified Fluid Mechanics Background

Since stable density stratification, in which lighter water overlies denser water, is a commonly observed situation in both the ocean and lakes, the density is often written as

$$\rho = \rho_0(\bar{\rho}(z) + \rho'(\vec{x}, t)) \tag{1.16}$$

where $\bar{\rho}(z)$ is referred to as the density stratification. For example, a smooth transition from fluid with $\rho = 1 - \Delta\rho$ to fluid with $\rho = 1 + \Delta\rho$ is given by

$$\bar{\rho}(z) = 1 + \Delta\rho \tanh\left(\frac{z - z_0}{d}\right)$$

where $z = z_0$ is the center of the transition region, and d sets its thickness. Here, $\Delta\rho$ is dimensionless, or in other words, the density change is specified as a fraction of ρ_0.

The decomposition (1.16) means that the density equation may be rewritten as

$$\frac{\partial \rho'}{\partial t} + \vec{u} \cdot \nabla \rho' + w \frac{d\bar{\rho}}{dz} = 0$$

or

$$\frac{D\rho'}{Dt} = -w \frac{d\bar{\rho}}{dz} \tag{1.17}$$

Similarly, the vorticity equation (in two dimensions for simplicity) can be rewritten as

$$\frac{D\omega}{Dt} = \frac{g}{\rho_0} \left(-\frac{\partial \rho'}{\partial y}, \frac{\partial \rho'}{\partial x} \right),$$

demonstrating that it is only the changes from the background stratification, and not the stratification itself, that lead to vorticity generation.

Since lines of constant density, the so-called isopycnals, run perpendicular to the direction of gravity when at rest, density stratification provides a wave guide for gravity waves in the interior of a fluid or internal waves. While the bulk of this book is about theories of internal waves, a very simple version of the simple harmonic oscillator for internal waves can be derived as follows. Consider a fluid particle at a height z_0. At z_0, the fluid has a density of $\bar{\rho}(z_0)$, and this is the density of the fluid particle. If the particle is raised a very small distance so it is now located at $z_0 + \Delta z$, the ambient density is $\bar{\rho}(z_0 + \Delta z) \neq \bar{\rho}(z_0)$. In the simplest situation, the only force per unit volume is the buoyancy, or the difference between the weight of the fluid particle, and the weight of the fluid it displaces (i.e., Archimedes' Principle). According to Newton's second law, this must be balanced by the rate of change with time of the linear momentum. Since the net force is non-zero, the fluid particle will move. Hence, we can write Newton's second law in equation form for a time dependent $\Delta z(t)$ as

$$\rho_0 \frac{d^2 \Delta z}{dt^2} = g(\bar{\rho}(z_0 + \Delta z) - \bar{\rho}(z_0)) \approx g\bar{\rho}'(z_0)\Delta z$$

where I used a Taylor series in the right most expression and the prime to denote a derivative. If we define

$$N^2(z) = -g \frac{d\bar{\rho}}{dz} \tag{1.18}$$

the equation may be rewritten as follows:

$$\frac{d^2 \Delta z}{dt^2} + N^2(z_0)\Delta z = 0. \tag{1.19}$$

We can immediately recognize $N^2(z_0)$ as the frequency squared from the simple harmonic oscillator equation. This frequency may change with the initial height of the particle but is otherwise held constant (this is valid since we assumed the displacement to be very small). The square root of $N^2(z_0)$ is referred to as the buoyancy frequency, or the Brünt–Väisälä frequency. Without the Boussinesq approximation, N^2 is defined with a factor of $1/\rho$, and some authors maintain a factor of $1/\rho_0$ even with the Boussinesq approximation. This, however, assumes that $\bar\rho(z)$ has the dimensions of $kg\ m^{-3}$, and throughout this book, I prefer to have $\bar\rho(z)$ to be dimensionless, specifying that fraction of density change associated with the stratification.

In the ocean, the stratification is often seen to have a consistent form. Moving downward, one observes: a nearly constant density in the surface layer (often called the surface mixed layer), a sharp decrease in density in what is called the **pycnocline**, and beneath this a much smaller decrease in density over the remainder of the water column. Laboratory experiments typically employ a similar stratification profile but ignore the stratification in the deep. The thickness of pycnocline in the laboratory is often, though not always, much thinner (when scaled by the total depth) than that observed in the ocean. Examples of $\bar\rho(z)$ and $N^2(z)$ for a lab-like and ocean-like stratification are shown in Fig. 1.3.

There is one final quantity that is defined using the background density profile. Say we measure the density at a location, (x, z). Consider a measurement of the density at a location, (x, z). If we then follow the isopycnal passing through this point to a location far upstream from any waves or other disturbances, and label this location by (x', z') we could then define $\eta(x, z) = z - z'$. In words, η is the vertical displacement of the isopycnal at (x, z) from its "rest" location due to any local disturbances. Mathematically, we would write the implicit equation

$$\rho(x, z) = \bar\rho(z - \eta). \tag{1.20}$$

Thus, η, the isopycnal displacement with units of length, provides a way to define "wave amplitude" of internal waves.

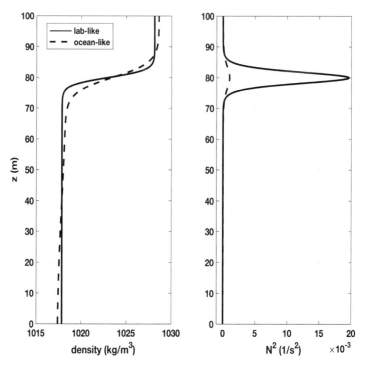

Fig. 1.3 Left panel: sample $\bar{\rho}(z)$ profiles for a lab-like (solid) and coastal ocean-like (dashed) stratification. Right panel: $N^2(z)$ for the same two profiles. Note how much smaller the peak in N^2 is in the broader, ocean-like stratification

Equation Boxes

BOX 1: Governing Equations for a Stratified, Inviscid Fluid:

1. The Boussinesq approximation holds.
2. The fluid is incompressible and inviscid.
3. Rotation can be neglected.

$$\rho_0 \frac{D\vec{u}}{Dt} = -\nabla p - \rho g \hat{k} \qquad \text{Conservation of Momentum}$$

$$\frac{D\rho}{Dt} = 0 \qquad \text{Conservation of Energy}$$

$$\nabla \cdot \vec{u} = 0 \qquad \text{Conservation of Mass}$$

BOX 2: Governing Equations Streamfunction-Density Form:

$$(u, w) = (\psi_z, -\psi_x) \qquad \text{Streamfunction}$$

$$\rho = \bar{\rho}(z) + \rho'(x, z, t) \qquad \text{density}$$

$$\omega = u_z - w_x = \nabla^2 \psi \qquad \text{vorticity}$$

$$\frac{D\nabla^2 \psi}{Dt} = \rho'_x g \qquad \text{Conservation of Vorticity}$$

$$\frac{D\rho'}{Dt} + w\bar{\rho}_z = 0 \qquad \text{Conservation of Energy}$$

BOX 3: General Linear Theory:

1. The wave amplitude is small enough so that nonlinear terms may be neglected

$$\psi_{TOTAL}(x, z, t) = \bar{\psi}(z) + \psi(x, z, t) \text{ where } \bar{\psi}_z = U(z) \qquad \text{streamfunction}$$

$$\psi(x, z, t) = \exp[i(kx - \sigma t)]\phi(z) \qquad \text{wave ansatz}$$

$$\phi_{zz} + \left(\frac{N^2(z)}{(c_0 - U)^2} + \frac{U''}{c_0 - U} - k^2 \right) \phi = 0 \qquad \text{vertical eigenvalue problem}$$

$$\phi(0) = \phi(H) = 0 \qquad \text{boundary conditions}$$

BOX 4: Simplest Linear Theory

1. The wave amplitude is small enough so that nonlinear terms may be neglected.
2. $U(z)=0$.

$$\psi(x, z, t) = \exp[i(kx - \sigma t)]\phi(z) \qquad \text{wave ansatz}$$

$$\phi_{zz} + \left(\frac{N^2(z)}{c_0^2} - k^2 \right) \phi = 0 \qquad \text{vertical eigenvalue problem (no U(z))}$$

$$\phi(0) = \phi(H) = 0 \qquad \text{boundary conditions}$$

BOX 5: Weakly Nonlinear Evolution Equations (WNL)

1. A perturbation expansion is performed in terms of amplitude (nonlinear terms) and aspect ratio (dispersive terms).
2. We separate out the vertical dependence like $\psi(x, z, t) = B(x, t)\phi(z)$.
3. We truncate to first order in the nonlinear and dispersive terms (KdV).
4. We truncate to first order in the nonlinear and dispersive terms but improve linear dispersion (BBM).
5. We truncate to second order in the nonlinear term and first order in the dispersive term (mKdV or Gardner).

$$B_t = -c_0 B_x \qquad\qquad\qquad\qquad\qquad\qquad\qquad \text{Linear Advection}$$

$$B_t = -c_0 B_x - \frac{\beta}{c} B_{xxt} \qquad\qquad\qquad\qquad\qquad \text{Linear Dispersion}$$

$$B_t = -c_0 B_x + \alpha B B_x + \beta B_{xxx} \qquad\qquad\qquad\qquad \text{KdV}$$

$$B_t = -c_0 B_x + \alpha B B_x - \frac{\beta}{c} B_{xxt} \qquad\qquad\qquad\quad \text{BBM}$$

$$B_t = -c_0 B_x + \alpha B B_x + \alpha_2 B^2 B_x + \beta B_{xxx} \qquad\quad \text{Gardner or mKdV}$$

$$B_t = -c_0 B_x + \alpha B B_x + \alpha_2 B^2 B_x - \frac{\beta}{c_0} B_{xxt} \qquad \text{Gardner BBM or mBBM}$$

BOX 6: Two Layer Fluid: Linear Solution, WNL Coefficients Parameters: $(h_1, h_2, \Delta\rho)$ upper and lower layer thicknesses and density jump, where $H(.)$ is the Heaviside step function

$$\phi(z) = z/h_1 + H(z - h_1)\,[-z/h_1 + 1 - (z - h_1)/h_2] \qquad \text{Eigenfunction}$$

$$c_0 = \sqrt{g\Delta\rho h_1 h_2/(h_1 + h_2)} \qquad\qquad\qquad\qquad \text{linear longwave speed}$$

$$\alpha = \frac{3}{2}c_0\frac{h_1 - h_2}{h_1 h_2} \qquad\qquad\qquad\qquad\qquad\qquad \text{KdV nonlinear parameter}$$

$$\beta = \frac{1}{6}c_0 h_1 h_2 \qquad\qquad\qquad\qquad\qquad\qquad\qquad \text{KdV dispersion parameter}$$

$$\alpha_2 = \frac{3c_0}{(h_1 h_2)^2}\left[\frac{7}{8}(h_1 - h_2)^2 - \left(\frac{h_1^3 + h_2^3}{h_1 + h_2}\right)\right] \qquad \text{Gardner nonlinear parameter}$$

BOX 7: WNL Solitary Wave Solution

$B(x,t) = B_0\mathrm{sech}^2(\gamma[x - ct])$	Solution form KdV
$B(x,t) = \dfrac{B_0}{b + (1 - b)\cosh^2(\gamma[x - ct])}$	Solution form Gardner
$c = c_0 + \dfrac{B_0}{3}\left(\alpha + \dfrac{1}{2}\alpha_2 B_0\right)$	WNL propagation speed
$\gamma^2 = \dfrac{B_0\left(\alpha + \frac{1}{2}\alpha_2 B_0\right)}{12\beta}$	Gardner width parameter
$\gamma^2 = \dfrac{B_0\alpha}{12\beta}$	KdV width parameter
$b = -\dfrac{B_0\alpha_2}{2\alpha + \alpha_2 B_0}$	Gardner weight parameter

---

BOX 8: DJL Equation

$\psi^T(x,z) = \psi^b(z) + \psi(x,z);\ \psi_z^b = V(z) = U(z) - c$	General streamfunction
$\psi^T(x,z) = \psi^b(z - \eta)$	Streamfunction in terms of η
$\psi(x,z) = -cz + c\eta$	Streamfunction in terms of η, no $U(z)$
$0 = \nabla^2\eta + \dfrac{N^2(z - \eta)}{c^2}\eta$	DJL equation no $U(z)$
$0 = \nabla^2\eta + \dfrac{N^2(z - \eta)}{(U(z - \eta) - c)^2}\eta + S_u$	full DJL equation
$S_u = \dfrac{U_z(z - \eta)}{U(z - \eta) - c}\left[1 - \left(\eta_x^2 + (1 - \eta_z)^2\right)\right]$	S_u for full DJL equation
$0 = \eta(x,0) = \eta(x,H);\ \eta \to 0$ as $|x| \to \infty$	DJL boundary conditions

Literature 1

Making suggestions for books on mathematical background is a very subjective matter. Almost all practitioners will have their "go to," and I have listed a few of mine already at the end of the Preface. For this chapter, I thus stick to a single, broad reference, and once "classical" textbook.

5. Boas, M.L. (2006) *Mathematical Methods in the Physical Sciences*, John Wiley. This is a book of mathematical methods and comes recommended from a colleague who has taught such material out of a number of sources. It has the advantage of being quite comprehensive in terms of material.

6. Whitham, G. B. (1974) *Linear and Nonlinear Waves*, John Wiley. The standard reference on the theory of dispersive waves and how they contrast with hyperbolic waves. Still a good read.

Chapter 2
Derivations: Linear, Weakly Nonlinear, and Conjugate Flow Theory

Modern software allows for the solution of the governing equations in full. The level of detail this is possible at (i.e., the numerical resolution) varies with access to high performance computation (HPC) facilities. Even if one solves the equations in full, data handling, data analysis and graphics are all in themselves a large topic. Historically, however, we have not had access to computation and hence a variety of techniques were developed to reduce the governing equations to simpler forms. In many cases these techniques took on a life of their own, and at times the original problems to be solved were lost completely. Since each generation learns the tools appropriate for the problems they are incentivized to study, it is probably fair to say that fewer practitioners of coastal oceanography today have full facility with the entire mathematical toolbox available to past generations.

Thus, this chapter aims to introduce three types of reductions of the full stratified Euler equations. Each leads to a widely used equation, or set of equations, and each has particular techniques associated with it that are useful to know about. We begin with linear theory, then focus on weakly nonlinear, or perturbation theory, and finally consider one flavor of fully nonlinear theory. The weakly nonlinear theory portion is split into two parts: (i) non-dimensionalization, which introduces the small parameters; (ii) expansion, which carries out the perturbation expansion that derives the KdV equation. For readers eager to get at the practical problems promised at the outset of the book, this chapter may be skipped on first reading.

The Taylor–Goldstein Equation

We begin with a background density profile $\bar{\rho}(z)$, which implies our fluid is stratified. We do not necessarily assume the stratification is stable, or in other words that $\bar{\rho}'(z) \leq 0$. If we consider a simple shear flow $\vec{u} = (U(z), 0)$, where for simplicity we will work in two dimensions, a simple substitution into the

M. Stastna, *Internal Waves in the Ocean*, Surveys and Tutorials in the Applied Mathematical Sciences 9, https://doi.org/10.1007/978-3-030-99210-1_2

stratified Euler equations (Box 1) shows that any $U(z)$ profile is a solution. This seems strange. Surely Nature has a way to choose which solutions may actually be realized! One clue comes from Continuum Mechanics, where it is shown that the central assumption behind the Euler equations is that the stress tensor in an inviscid fluid is isotropic, or has no preferred direction. A mathematical theorem then implies that the stress is of the form $-p\mathbf{I}$ and that there is no way for motion in the x direction, for example, to transmit forces in the z direction. The Navier–Stokes equations, which govern viscous flow, do have a means for motion in the x direction, for example, to transmit forces in the z direction. Thus in the "real world" viscosity places some limitations on the form $U(z)$ can take.

There is, however, a much larger point of view, which finds its expression in the theory of hydrodynamic stability. Here the idea is to take the base state $(u, w, \rho) = (U(z), 0, \bar{\rho}(z))$ and subject it to small perturbations and find an equation that governs where the perturbations grow or shrink. The classical theory of wave motion is then a subset of this theory that occurs when perturbations neither shrink nor grow (sometimes called "neutral" modes). While the main goal of this section is to derive the so-called Taylor–Goldstein equation, an ulterior motive is to provide an example of hydrodynamic stability theory.

Let us take the background velocity to be $(U(z), 0)$ and define the streamfunction from the perturbation velocity so that $(u, w) = (\psi_z, -\psi_x)$, with perturbation vorticity given by $\omega' = \nabla^2 \psi$. The equations in Box 2 can be linearized to read

$$\rho'_t + U\rho'_x + \psi_x \frac{N^2}{g} = 0, \tag{2.1a}$$

$$\nabla^2 \psi_t + U\nabla^2 \psi_x - \psi_x U_{zz} = \rho'_x g, \tag{2.1b}$$

where $N^2(z)$ is again the squared buoyancy frequency. We assume that the small perturbations discussed above have the formal form of plane wave solutions of the form

$$\rho'(x, z, t) = \hat{d}(z)e^{ik(x-ct)}, \tag{2.2a}$$

$$\psi(x, z, t) = \phi(z)e^{ik(x-ct)}, \tag{2.2b}$$

where $\phi(z)$ is the vertical structure of the streamfunction, and $d(z)$ is the vertical structure of the density perturbation. Notice that if perturbations grow we must have that either k or c have an imaginary part. Substitution and some algebraic simplification then give the set of equations

$$-c\hat{d} + \hat{d}U + \phi\frac{N^2}{g} = 0, \tag{2.3a}$$

$$-c(\phi_{zz} - k^2\phi) + U(\phi_{zz} - k^2\phi) - U_{zz}\phi = \hat{d}g. \tag{2.3b}$$

The first of these two equations can be further rearranged as

$$\hat{d}g = -\frac{\phi N^2}{U - c}.$$

Substitution of this relation into the second equation and rearrangement give

$$(U - c)(\phi_{zz} - k^2\phi) - U_{zz}\phi = -\frac{\phi N^2}{U - c}$$

which upon moving all unknowns to one side yields what is called the Taylor–Goldstein, or T-G, equation

$$\phi_{zz} + \left(\frac{N^2(z)}{(U - c)^2} - \frac{U''}{U - c} - k^2\right)\phi = 0. \tag{2.4}$$

Since the flow is inviscid we have $w = -\psi_x = 0$ at the boundaries, which means $\phi(0) = \phi(H) = 0$.

The T-G is admittedly very nice and compact (for a discussion in the context of hydrodynamic stability see [1]). It however has a number of annoying features as well: (i) there are few exact solutions, especially when $U(z) \neq 0$, (ii) it does not reduce to a standard eigenvalue problem when it is discretized. We will discuss the second issue in the next chapter, but the first merits at least some consideration here.

Let us begin with the simplest case; that of waves in a linear stratification with no shear and in the longwave, or $k \to 0$ limit. We take

$$\bar{\rho}(z) = 1 - \Delta\rho\frac{z}{H}$$

so that

$$N^2(z) = \frac{g\Delta\rho}{H} := N_0^2$$

and the T-G equation reduces to

$$\phi_{zz} + \frac{N_0^2}{c^2}\phi = 0.$$

It is clear that this is analogous to the simple harmonic oscillator with a general solution (I have chosen R_i for the two arbitrary constants since c is reserved for the eigenvalue)

$$\phi(z) = R_1 \sin(\frac{N_0}{c}z) + R_2 \cos(\frac{N_0}{c}z).$$

The bottom boundary condition $\phi(z = 0) = 0$ means $R_2 = 0$ and the upper boundary condition at $z = H$ implies

$$\sin(\frac{N_0}{c} H) = 0,$$

which means $N_0 H / c = n\pi$ for $n = 1, 2, \ldots$ Solving for c and using the expression to simplify gives the final solution (remember that for a linear problem $R_1 = 1$ is a reasonable choice),

$$\phi_n(z) = \sin\left(\frac{n\pi}{H} z\right) \tag{2.5}$$

$$c_n = \frac{N_0 H}{n\pi}. \tag{2.6}$$

There are several points about the solution that can be immediately noted: (i) $c_1 > c_2 > c_3 > \ldots$, (ii) waves have a sinusoidal structure, with more oscillations as n increases. In particular, $n = 1$ implies that the eigenfunction is one half of a sinusoid's period, $n = 2$ implies that the eigenfunction is one sinusoidal period and so on.

Since waves are the main topic of multiple chapters that will follow, let us consider something different, namely the problem of an unstable stratification. Since we aim to do everything analytically let us again take a very special $\bar{\rho}(z)$ profile, namely

$$\bar{\rho}(z) = 1 + \alpha z \tag{2.7}$$

where $\alpha > 0$. The buoyancy frequency squared in this case is constant, but negative,

$$N^2(z) = -g\alpha$$

and the stratification has denser water overlying lighter water. For anyone who has ever turned a glass upside down, you know that this situation cannot be stable! Let us see how the mathematics reflects this. The T-G equation reduces to

$$\phi_{zz} + \left(-\frac{g\alpha}{c^2} - k^2\right)\phi = \phi_{zz} - \lambda^2\phi = 0.$$

In the second step I just factored out a (-1) and grouped together everything in the brackets, so that

$$\lambda^2 = \frac{g\alpha}{c^2} + k^2.$$

The solutions are exponentials, $\phi = R_1 \exp(\lambda z) + R_2 \exp(-\lambda z)$. This seems like a disaster, as we have no way to satisfy the two boundary conditions and still have a nontrivial solution. But this assumed c was a real number. What if it is an imaginary number, $c = iM$? In that case we have

$$\phi_{zz} + \left(\frac{\alpha g}{M^2} - k^2\right)\phi = 0$$

and now we can have sinusoidal solutions provided

$$\frac{\alpha g}{M^2} - k^2 > 0.$$

Since we boundary conditions of $\phi(0) = \phi(H) = 0$ let us try

$$\phi = \sin(mz).$$

$\phi(H) = 0$ implies $m_n = n\pi/H$ where $n = 1, 2, 3\ldots$ and the T-G equation reduces to

$$-m_n^2 + \frac{\alpha g}{M^2} - k^2 = 0.$$

This allows us to solve for M as function of (k, m) and n:

$$M_n^2 = \frac{\alpha g}{k^2 + m_n^2}.$$

If we return to the form of the streamfunction we see that

$$\psi = \exp[ik(x - iM_n t)]\sin(m_n z) = \exp(ikx)\exp(kM_n t)\sin(m_n z)$$

where we have taken the positive root for M_n. This implies that in this case the perturbation grows and we can define a growth time as

$$T_{growth} = \frac{1}{kM_n} = \sqrt{\frac{k^2 H^2 + n^2\pi^2}{\alpha g k^2 H^2}}.$$

This is a tidy expression and allows us to conclude a few things: (i) Since α tells us how fast density increases with height, we see a larger density increase leads to a faster growth rate. (ii) n tells us how many oscillations we have in the vertical and we can see that with more oscillations we have a slower growth rate. There are also a few things that the analytical theory is not great for that we should note: (i) It is restrictive ($U(z) = 0$, $\bar{\rho}(z)$ has to have a special form), (ii) it is not very visual, (iii) the theory does not tell us what it is limitations are—e.g., when should we be worried that we dropped nonlinear terms? This is what will eventually lead us to numerical methods. But first let us consider where it led people to historically.

Non-dimensionalization

In the previous section we derived linear theory for disturbances of a stratified shear flow. We did this by dropping the nonlinear terms in the governing equations and did so in a more or less *ad hoc* manner. In the next two sections we want to do this in a more systematic manner via a non-dimensionalization of the governing equation, which will yield two small parameters, and a perturbation expansion in those two small parameters. The first small parameter will be the amplitude of the disturbance, and in this sense will be just like the previous section EXCEPT we will not just calculate the leading order approximation but its first correction as well. The second parameter is motivated by observations that suggest that the horizontal length scale, L, of energetic internal waves in the sea is much larger than the typical vertical scale (say the total depth). The small parameter is the aspect ratio $H/L \ll 1$ and we will expand in the square of this parameter, or in terms of

$$\mu = \left(\frac{H}{L}\right)^2 .$$

A typical situation is shown in Fig. 2.1, along with a picture of what internal waves "really look like." The latter is shown because when presenting the results of simulations visually it is natural to distort the axes.

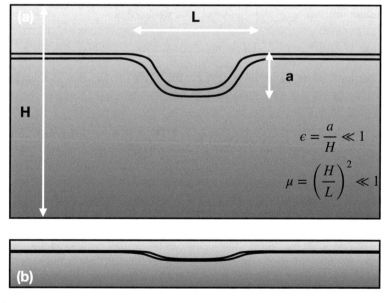

Fig. 2.1 (**a**) Schematic of internal wave motion showing where the two small parameters of weakly nonlinear KdV theory come from. (**b**) A reminder of the "actual" appearance of internal wave since the axes are often distorted in visualizations of simulations

We consider the two dimensional stratified Euler equations written in terms of the streamfunction ψ and using the bilinear operator $J[a, b]$. Cross-differentiating the two momentum equations and subtracting give the vorticity equation, which may be written as

$$\frac{D\nabla^2 \psi}{Dt} = \rho'_x g$$

where we have also split the density into a background stratification $\bar{\rho}(z)$ and a perturbation $\rho'(x, z, t)$. Introducing the variable

$$b = g\rho' \tag{2.8}$$

the vorticity equation may be rewritten with all the nonlinear terms on one side (and using subscripts to denote partial derivatives) as

$$\left(\nabla^2 \psi\right)_t - b_x = J[\psi, \nabla^2 \psi]. \tag{2.9}$$

The right hand side takes this because

$$J[\psi, \nabla^2 \psi] = \psi_x (\nabla^2 \psi)_z - \psi_z (\nabla^2 \psi)_x = -u(\nabla^2 \psi)_x - w(\nabla^2 \psi)_z$$

or in other words, because it is a re-expression of the nonlinear terms of the material derivative. The density equation reads

$$\frac{D\rho'}{Dt} + w\bar{\rho}_z = 0$$

and multiplying by g this can be rewritten as

$$b_t + N^2(z)\psi_x = J[\psi, b]. \tag{2.10}$$

We assume $N^2(z) \geq 0$, or that the fluid is stably stratified. Next non-dimensionalize (tildes denote dimensionless variables) the spatial dimensions as

$$(x, z) = (L\tilde{x}, H\tilde{z}).$$

Since N^2 has the dimensions of frequency squared, it provides a natural time scale. Simply write

$$N(z) = N_0 \tilde{N}(\tilde{z})$$

a characteristic wave speed is thus $c_0 = H N_0$ and a characteristic time scale is given by

$$t = \frac{L}{N_0 H} \tilde{t}.$$

The streamfunction is scaled so that its amplitude, ϵ, is our first small parameter. ϵ is referred to as the nonlinearity parameter, and ψ is scaled as

$$\psi = \epsilon \Psi \tilde{\psi}.$$

Since we have already chosen length and time scales we must determine Ψ in terms of existing parameters. Recall $u = \psi_z$ and a bit of algebra gives

$$u = \epsilon \frac{\Psi}{H} \tilde{\psi}_z$$

and thus we should choose

$$\Psi = c_0 H.$$

We can check that this is a sensible choice by using the conservation of mass $u_x + w_z = 0$. We can write the left hand side of this expression using the non-dimensionalizations we have as

$$\epsilon \frac{c_0 H}{H L} \tilde{\psi}_{zx} - \epsilon \frac{c_0 H}{L H} \tilde{\psi}_{xz}$$

and we are looking to ensure that both terms are of the same relative size, which an inspection reveals to be true. The final variable to scale is b, which we can formally write as

$$b = \epsilon B \tilde{b}.$$

Because the density perturbations are small (but not infinitesimal), it makes sense that b is proportional to ϵ. Now recall that we wrote the b equation with the linear terms on one side and the quadratic terms on the other. This means one way to find B is to assume the two terms on the left hand side are about the same size. This means

$$\frac{B}{T} = N_0^2 \frac{\Psi}{L}$$

or

$$B = \frac{L}{c_0} N_0^2 c_0 H \frac{1}{L} = N_0^2 H$$

so that

$$b = \epsilon N_0^2 H \tilde{b}.$$

Finally we need to consider the effect of the discrepancy in length on partial derivatives

$$\nabla^2 = \frac{1}{L^2}\partial_{\tilde{x}\tilde{x}} + \frac{1}{H^2}\partial_{\tilde{z}\tilde{z}} = \frac{1}{H^2}\left(\frac{H^2}{L^2}\partial_{\tilde{x}\tilde{x}} + \partial_{\tilde{z}\tilde{z}}\right) = \frac{1}{H^2}\left(\mu\partial_{\tilde{x}\tilde{x}} + \partial_{\tilde{z}\tilde{z}}\right).$$

Substituting all this information, and dropping the tildes, allows us to write the dimensionless equations as

$$\psi_{zzt} - b_x = \epsilon J[\psi, \psi_{zz}] - \mu\psi_{xxt} + \epsilon\mu J[\psi, \psi_x x] \tag{2.11}$$

$$b_t + N^2(z)\psi_x = \epsilon J[\psi, b]. \tag{2.12}$$

KdV Theory

The discussion in this section essentially follows [7], with a minor tweak of notation. We now seek a perturbation expansion in the two small parameters (h.o.t. stands for "higher order terms")

$$\psi = \psi^{(0)} + \epsilon\psi^{(1,0)} + \mu\psi^{(0,1)} + \text{h.o.t.}$$

$$b = b^{(0)} + \epsilon b^{(1,0)} + \mu b^{(0,1)} + \text{h.o.t.} \tag{2.13}$$

Since the procedure involves a fair bit of algebra, let us make sure at the outset that we are clear on what we want to happen. We want to start with the leading order problem and find a linear equation of wave type (we could even hope for the classical wave equation). We then seek to correct this wave equation with terms that are $O(\epsilon)$ and terms that are $O(\mu)$. Let us see how that works out.

The $O(1)$ Problem

The equations governing the leading order problem are a simplified form of the linearized equations (remember we are also dropping corrections in the aspect ratio),

$$\left(\psi_{zz}^{(0)}\right)_t - b_x^{(0)} = 0$$

$$b_t^{(0)} + N^2(z)\psi_x^{(0)} = 0. \tag{2.14}$$

A simple cross-differentiation and addition allows us to derive a single equation for $\psi^{(0)}$,

$$\psi_{zztt}^{(0)} + N^2(z)\psi_{xx}^{(0)} = 0$$

which looks a little bit like a wave equation from classical PDEs, except for the z derivatives. We posit a separable solution

$$\psi^{(0)} = c_0 B(x, t)\phi(z)$$

and a standard separation of variables procedure yields

$$\frac{B_{tt}}{B_{xx}} = -N^2(z)\frac{\phi}{\phi_{zz}}.$$

Choosing the separation constant to be c_0^2 allows us to recover the classical wave equation for B, namely

$$B_{tt} = c_0^2 B_{xx}. \tag{2.15}$$

The vertical structure is governed by the equation

$$\phi_{zz} + \frac{N^2(z)}{c_0^2}\phi = 0. \tag{2.16}$$

The boundary conditions at the flat bottom and rigid upper boundary read

$$\phi(0) = \phi(1) = 0.$$

Together these are an eigenvalue problem of Sturm–Liouville type (with $p(z) = 1$ and $\sigma(z) = N^2(z)$) for the propagation speed c (the eigenvalue in Sturm–Liouville form would be $\lambda = 1/c^2$) and the vertical structure function $\phi(z)$. Since the stratification is stable we have $N^2(z) \geq 0$ and we can mine Sturm–Liouville theory for some facts about the eigenvalue problem: (i) There are an infinite number of solutions $(\phi_n(z), c_n)$ with $c_n^2 > 0$. (ii) The eigenvalues may be ordered so that the positive roots, or rightward propagation speeds satisfy, $c_1 > c_2 > c_3 > \ldots > 0$. We will consider rightward propagating waves only, for the remainder of the derivation (leftward propagating waves are considered in one of the mini-projects). As an aside, the inner product induced by our particular Sturm–Liouville problem is

$$\langle f, g \rangle = \int_0^1 f(z)g(z)N^2(z)dz$$

though in the work below we will use the standard inner product.

There are two outstanding issues. The first is purely algebraic; we need to find an expression for $b^{(0)}$. From the equation for b we have

$$b_t^{(0)} = -N^2(z)c_0 B_x(x, t)\phi(z)$$

but we know that for rightward propagating waves $B_t = -c_0 B_x$ so that

$$b_t^{(0)} = N^2(z)B_t(x, t)\phi(z)$$

Thus we can write $b^{(0)}$ as

$$b^{(0)} = B(x, t)N^2(z)\phi(z).$$

The second outstanding issue is conceptual. The linear theory is the first step in a perturbation expansion, so how should we represent mathematically the physics that is missing at leading order? Put another way, we only know that $B_t = -cB_x$ at leading order. It thus seems like a good idea to represent our lack of knowledge at higher orders explicitly. We do this by writing

$$B_t = -c_0 B_x - \epsilon R(x, t) - \mu Q(x, t) + \text{h.o.t.} \tag{2.17}$$

where we expect this to hold in the formal limit $\epsilon, \mu \to 0$. At this point in time we do not know what $R(x, t)$ and $Q(x, t)$ are, but we hope to find them as part of the expansion. What these "correction" terms are meant to represent physically is shown in Fig. 2.2.

Returning to the expansion, if we substitute (2.17) into the leading order equations we see that the first equation has a contribution

$$-\epsilon R(x, t)\phi_{zz} - \mu Q(x, t)\phi_{zz}$$

on the left hand side, while the second equation has a contribution

$$-\epsilon R(x, t)N^2(z)\frac{\phi}{c_0} - \mu Q(x, t)N^2(z)\frac{\phi}{c_0}$$

on the left hand side. These terms will contribute at $O(\epsilon)$ and $O(\mu)$.

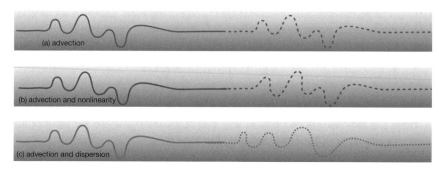

Fig. 2.2 The conceptual idea of the physical processes the corrections in WNL are meant to represent (**a**) The uncorrected case: advection only. (**b**) Advection and nonlinearity, some steepening is visible, (**c**) Advection and dispersion: longer waves are seen to propagate faster

The $O(\epsilon)$ Problem

The governing equations read

$$-c_0 R(x,t)\phi_{zz} + \psi_{zzt}^{(1,0)} - b_x^{(1,0)} = J[\psi^{(0)}, \psi_{zz}^{(0)}]$$

$$-R(x,t)N^2(z)\phi + b_t^{(1,0)} + N^2(z)\psi_x^{(1,0)} = J[\psi^{(0)}, b^{(0)}]. \qquad (2.18)$$

Since the RHS consists only of known functions let us work it out first

$$J[\psi^{(0)}, \psi_{zz}^{(0)}] = c_0^2 B B_x(\phi\phi_{zzz} - \phi_z\phi_{zz})$$

and

$$J[\psi^{(0)}, b^{(0)}] = -c_0^3 B B_x(\phi\phi_{zzz} - \phi_z\phi_{zz}).$$

These are insightful because they suggest that the (x,t) dependence of all terms should have the form $B B_x$. This implies a very sensible guess would be to try

$$R(x,t) = \alpha B B_x \qquad (2.19)$$

where α is our "wiggle room" and we make a mental note that we should determine α from the theory. To get a factor of $B B_x$ from the other terms we make the inspired guess (we are really following the leading order guess)

$$\psi^{(1,0)} = c_0^2 B(x,t)^2 \phi^{(1,0)}(z)$$
$$b^{(1,0)} = B(x,t)^2 N^2(z) d^{(1,0)}(z). \tag{2.20}$$

Substitution gives

$$-c_0 \alpha B B_x \phi_{zz} + 2c_0^2 B B_t \phi_{zz}^{(1,0)} - 2 B B_x N^2(z) d^{(1,0)} = c_0^2 B B_x (\phi\phi_{zzz} - \phi_z\phi_{zz})$$
$$-\alpha B B_x N^2(z)\phi + 2 B B_t N^2(z) d^{(1,0)} + N^2(z) c_0^2 2 B B_x \phi^{(1,0)} = -c_0^3 B B_x (\phi\phi_{zzz} - \phi_z\phi_{zz}). \tag{2.21}$$

We know that $B_t = -c_0 B_x + O(\epsilon) + O(\mu)$ so that on multiplying the first equation by c_0 we get

$$-c_0^2 \alpha B B_x \phi_{zz} - 2c_0^4 B B_x \phi_{zz}^{(1,0)} - 2c_0 B B_x N^2(z) d^{(1,0)} = c_0^3 B B_x (\phi\phi_{zzz} - \phi_z\phi_{zz})$$
$$-\alpha B B_x N^2(z)\phi - 2c_0 B B_x N^2(z) d^{(1,0)} + N^2(z) c_0^2 2 B B_x \phi^{(1,0)} = -c_0^3 B B_x (\phi\phi_{zzz} - \phi_z\phi_{zz}). \tag{2.22}$$

we can now cancel the $B B_x$ common factor and subtract the two equations to eliminate the two $d^{(1,0)}$ terms

$$\alpha(-c_0^2 \phi_{zz} + N^2(z)\phi) - 2c_0^4 \phi_{zz}^{(1,0)} - N^2(z) c_0^2 2\phi^{(1,0)} = 2c_0^3 B B_x (\phi\phi_{zzz} - \phi_z\phi_{zz}).$$

Next note that $-c_0^2 \phi_{zz} = N^2(z)\phi$, eliminate the common factor of 2, and divide by c_0^2 to find

$$\alpha\phi_{zz} + c_0^2 \left(\phi_{zz}^{(1,0)} + \frac{N^2(z)}{c_0^2}\phi^{(1,0)} \right) = -c_0(\phi\phi_{zzz} - \phi_z\phi_{zz})$$

and dividing by $2c_0^4$ and rearranging we finally get

$$c_0^2 \left(\phi_{zz}^{(1,0)} + \frac{N^2(z)}{c_0^2}\phi^{(1,0)} \right) = -\alpha\phi_{zz} - c_0(\phi\phi_{zzz} - \phi_z\phi_{zz}). \tag{2.23}$$

Since the physical boundary conditions are unchanged by the expansion we must have $\phi^{(1,0)}(1) = \phi^{(1,0)}(0) = 0$. However, since c_0 is known this makes (2.23) a boundary value problem. To ensure solvability of the BVP we invoke the Fredholm alternative of the Sturm–Liouville eigenvalue problem (2.16). In practice, this means multiplying (2.23) by ϕ and integrating over the domain (the "standard" inner product). The left hand side gives

$$\int_0^1 [\phi\phi_{zz}^{(1,0)} + \frac{N^2(z)}{c_0^2}\phi\phi^{(1,0)}]dz = \int_0^1 [\phi\phi_{zz}^{(1,0)} - \phi_{zz}\phi^{(1,0)}]dz$$

$$= \int_0^1 [-\phi_z\phi_z^{(1,0)} + \phi_z\phi_z^{(1,0)}]dz$$

$$= 0 \qquad\qquad (2.24)$$

where in the first line I used (2.16) and in the second line I integrated each term by parts. To simplify the right hand side we note that integration by parts shows that (details as an exercise)

$$\int_0^1 (\phi^2\phi_{zzz} - \phi\phi_z\phi_{zz})dz = \frac{3}{2}\int_0^1 (\phi_z)^3 dz.$$

With this fact in hand the right hand side simplifies to a formula for α, namely

$$\alpha = \frac{3}{2}c_0 \frac{\int_0^1 (\phi_z)^3 dz}{\int_0^1 (\phi_z)^2 dz}. \qquad\qquad (2.25)$$

The $O(\mu)$ Problem

Thankfully the algebra of the $O(\mu)$ problem is a bit easier. We will give fewer details.

$$-c_0 Q\phi_{zz} + \psi_{zzt}^{(0,1)} - b_x^{(0,1)} = \psi_{xxt}^{(0)} = c_0^2 B_{xxx}\phi$$

$$-QN^2\phi + b_t^{(0,1)} + N^2(z)\psi_x^{(0,1)} = 0$$

I have used the fact that $B_t = -c_0 B_x$ to leading order on the right hand side (higher order extensions would need much more care!). There is only one non-zero term on the right hand side and this suggests Q should have the form βB_{xxx} where β is again our "wiggle room." We can also see that both $\psi^{(0,1)}$ and $b^{(0,1)}$ should be proportional to B_{xx} and some fairly straight-forward algebra yields the first dispersive boundary value problem for the vertical structure (details left as an exercise)

$$\phi_{zz}^{(0,1)} + \frac{N^2(z)}{c_0^2}\phi^{(0,1)} = -\phi - \frac{2\beta}{c_0}\phi_{zz} \qquad\qquad (2.26)$$

where $\phi^{(0,1)}(1) = \phi^{(0,1)}(0) = 0$. The Fredholm Alternative gives the formula for the dispersive coefficient as

$$\beta = \frac{c_0}{2} \frac{\int_0^1 (\phi)^2 dz}{\int_0^1 (\phi_z)^2 dz}. \tag{2.27}$$

Comparing Eqs. (2.27) and (2.25) we note that $\beta \geq 0$ while α may take either sign.

Extension to higher order increases the complexity of the calculations, and the interested reader may pursue the details in references like [7] and [8].

Commentary

The primary job of the perturbation expansion was to derive the KdV equation (Box 5)

$$B_t = -c_0 B_x + \alpha B B_x + \beta B_{xxx}$$

for the horizontal and temporal evolution of waves. This would not be much use unless the method ALSO gave a way to find the coefficients α and β. While these were derived via the Fredholm alternative for Sturm–Liouville problems, the method of derivation does not really provide much physical meaning. But if we look at the Eqs. (2.27) and (2.25) we see these are in terms of ϕ, the solution of the leading order eigenvalue problem for the vertical structure. The eigenvalue problem has only one physical "parameter," the square of the buoyancy frequency $N^2(z)$. Thus α and β are implicit functions of the particular stratification in a given problem.

The derivation was presented as an attempt to lead the reader to "how we get to the solution." If one were interested in extending the method to higher order, a systematic choice of notation (see for example Lamb and Yan) would be better. The situation also essentially cries out for computer algebra. This is doubly so if we wished to include a background shear current $U(z)$. Nevertheless, at the time of writing, I am unaware of any attempt to do this.

Since the KdV equation is itself nonlinear, and in fact even the linear eigenvalue problem (2.16) has few solutions in closed form, the reader may wonder why even bother with such a complicated procedure. The first answer is historical. In the 1970s, well after the derivation of the KdV equation in the context of internal waves, solving the KdV proved a stern test for both the techniques and computational set ups available at that time (see the discussion in [3]). The second answer is conceptual; the KdV equation is the simplest equation that contains a tunable nonlinear term, $\alpha B B_x$, and a tunable dispersive term, βB_{xxx}. It thus allows for novel physics that balances the two. We will explore more in the following chapter.

Conjugate Flow Theory

Conjugate flow theory (following the reference [9]) makes a very different set of choices from the KdV theory described in the previous section. In Fig. 2.3 we demonstrate the basic set up. The idea is that a wave with a flat crest region for which the flow can be considered as purely horizontal connects in a smooth manner to a stably stratified region found far upstream ($x \rightarrow \infty$). The wave is propagating with some unknown speed, labelled c_j. The question is whether a simple equation (i.e., simpler than the stratified Euler equations) can be derived for this situation. The reader is forewarned that I feel obliged to give all the algebraic steps for this case.

We begin by demanding that all streamlines connect to $\pm\infty$. This precludes regions of overturning and allows us to restate the definition of isopycnal displacement (1.20) for this context as

$$\bar{\rho}_p(z) = \bar{\rho}(z - \eta).$$

It also allows us to state that the state within the wave $(U(z) - c_j, \bar{\rho}_p(z), \bar{p}_p(z))$ is connected to the state far upstream $(-c_j, \bar{\rho}(z - \eta), \bar{p}(z - \eta))$ by a streamline. Since streamlines cannot cross, and the streamline ahead of the wave is at a height $z - \eta(z)$ we must have that $z - \eta(z)$ is an increasing function of z. This immediately implies $1 - \eta_z(z) > 0$ or

$$\eta_z < 1,$$

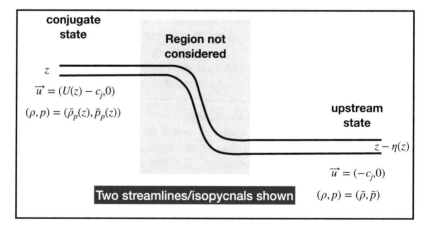

Fig. 2.3 Definition diagram for a conjugate flow. Far upstream a uniform, stably stratified flow is connected by a transition region that is not part of the theory to another purely horizontal flow that is referred to as "conjugate." See text for details

which is the condition for no vertical streamlines, or in other words the condition for no wave breaking. Imposing Conservation of Volume between any two streamlines gives (Exercise)

$$U = c_j \eta_z(z)$$

where the right hand side sets the volume flux far upstream. Next invoke Bernoulli's theorem under the Boussinesq approximation to write

$$\frac{1}{2}\rho_0 c_j^2 + \bar{p}(z - \eta) + g\bar{\rho}(z - \eta)(z - \eta) = \frac{1}{2}\rho_0 c_j^2(\eta_z - 1)^2 + \bar{p}_p(z) + g\bar{\rho}_p(z)z$$

here ρ_0 is a constant reference density. Next note that in both upstream and downstream the pressure is hydrostatic so that we have

$$\frac{d\bar{p}_p(z)}{dz} = -g\rho_0\bar{\rho}_p(z) = -g\rho_0\bar{\rho}(z - \eta)$$

and

$$\frac{d\bar{p}(z - \eta)}{dz} = -g\rho_0\bar{\rho}(z - \eta)(1 - \eta_z).$$

This means

$$\frac{d\bar{p}(z - \eta)}{dz} - \frac{d\bar{p}_p(z)}{dz} = g\rho_0\eta_z\bar{\rho}(z - \eta)$$

We now differentiate the expression from Bernoulli's theorem with respect to z and use the relation we just derived for the pressure derivatives to give

$$g\rho_0\eta_z\bar{\rho}(z-\eta) + g\rho_0\bar{\rho}(1-\eta_z) + g\rho_0\bar{\rho}'(z-\eta)(1-\eta_z)(z-\eta) = -\rho_0 c_j^2\eta_{zz}(1-\eta_z) + \rho_0 g\bar{\rho}(z-\eta) + gz\rho_0\bar{\rho}'(z-\eta)(1-\eta_z).$$

Cancelling ρ_0 and simplifying gives

$$g\bar{\rho}'(z - \eta)(1 - \eta_z)(z - \eta) = -c_j^2\eta_{zz}(1 - \eta_z) + gz\rho_0\bar{\rho}'(z - \eta)(1 - \eta_z)$$

which has a common factor of $1 - \eta_z$. Simplifying and rearranging give

$$\eta_{zz} - g\bar{\rho}'(z - \eta)\frac{1}{c_j^2}\eta = 0$$

which may be simplified further using the definition of the buoyancy frequency squared,

$$N^2(z - \eta) = -g\bar{\rho}'(z - \eta)$$

so that

$$\eta_{zz} + \frac{N^2(z - \eta)}{c_j^2}\eta = 0. \tag{2.28}$$

The boundary conditions state that there can be no isopycnal displacement across the boundary so that $\eta(z) = \eta(H) = 0$. This is a nonlinear ordinary differential eigenvalue problem, and an extraordinarily tidy one at that. There is some subtlety in solving it numerically (see [9]), but after all the algebra it must be admitted that on this occasion the theoretical manipulations did indeed yield a significant simplification!

Concluding Thoughts

This chapter has been all about theoretical tools. I have covered topics in linear, weakly nonlinear, and fully nonlinear theory. Part of the motivation was to introduce what is used later, but throughout I was very conscious of the fact that most textbook expositions of theory in fluid mechanics concentrate on linear theory. I provided an example of such an approach in deriving the Taylor–Goldstein equation and showing analytically what happens for an unstably stratified fluid with a very special density profile (i.e., the linearly stratified fluid). Here I used the so-called wave ansatz, which is a generalization of the plane wave solution of classical physics. Essentially if we write the complex exponential form for a horizontally propagating wave with a known, real wave number k we have

$$\exp(ik(x - ct)) = \exp(ik(x - c_R t))\exp(kc_I t)$$

and we can see that a complex c leads to possible growth (and hence instability) while a purely real c leads to wave motion.

For weakly nonlinear theory, we began by non-dimensionalizing the governing equations in order to establish unambiguous small parameters for our subsequent perturbation expansion. The goal was to modify the linear longwave theory to account for small, but finite amplitude and small, but finite aspect ratio (i.e., finite wavelength waves). The result was not only a governing equation (the KdV equation of Box 5), but an expression for the terms in this expression representing nonlinearity and dispersion. While these are commonly quoted in the literature, it is important to note that they depend on how one chooses to write the variables we are expanding. For example, in our expression for the streamfunction, ψ, we specifically chose to include a factor of c_0. The expressions for α and β then also have a factor of c_0. Had we not included c_0 in the expression for ψ the factor of c_0 would still occur in the expression for β, the dispersive coefficient, but would be absent in the expression for the nonlinear coefficient, α. As an aside, we note that the presence of a background shear may be included in the KdV theory, for example as in [10].

Finally we presented an example of an exact theory. Here the central lesson is that to successfully represent the full stratified Euler equations a rather drastic approximation must be made. For the conjugate flow theory presented in this chapter, that approximation states that we do seek ONLY the purely horizontal flow in the "conjugate" region. Thus the shape of a wave is beyond the scope of the theory. The theory introduces the notion of the isopycnal displacement η, and its use to simplify the governing equations. This will turn out to be important in the theory of the Dubreil Jacotin Long equation presented in Chap. 5.

There are theories that sit somewhere between exact, and weakly nonlinear. The reference [11] is a recent example of the use of such a theory, with some relevance to the material in Chap. 7.

As is typical of applied mathematical theory, this chapter featured considerable algebra. I tried to provide for the reader the necessary steps to see how one gets to the final expression, something that is often absent in the literature and even textbooks. Nevertheless, it must be said that the derivation cannot be the finish line for any scientifically minded reader and we must spend as much, or perhaps even more, time on learning to use the various aspects of the theory we have derived in this chapter.

Literature 2

7. K. G. Lamb and L. Yan (1996) The evolution of internal wave undular bores: comparisons of a fully nonlinear numerical model with weakly-nonlinear theory, J. Phys. Oceanogr., 26, 2712. This paper presents the version of weakly nonlinear theory (WNL) that I am following (with a minor switch of notation) in the above chapter. The paper also does something that many WNL papers do not; it compares with a simulation. Not surprisingly it finds that WNL is not great as a quantitative tool.

8. J. A. Gear and R. Grimshaw (1983) A second-order theory for solitary waves in shallow fluids, Phys. Fluids, 26, 14. Roger Grimshaw (with various co-workers) has covered every possible take on WNL. I chose an older paper to include here because it shows clearly the difficulty in algebra of extending to higher order.

9. Lamb, K.G., Wan, B. (1998) Conjugate flows and flat solitary waves for a continuously stratified fluid, Phys. Fluids, 10(8), 2061–2079. While the concept of conjugate flow goes back to the work of Benjamin in the 1960s, this paper provides a clear derivation in a manner that exposes the link to internal solitary waves (ISWs). It was also the first to compare flat crested ISWs to conjugate flows.

10. M. Stastna and K. G. Lamb, (2002) Large fully nonlinear internal solitary waves: The effect of background current, Phys. Fluids, 14, 2987. This paper presents the WNL with a background current (many others have done this), but more importantly demonstrates how the presence of a background shear changes exact internal solitary waves. As such I choose to list it since I can point to it for further topics in subsequent Chapters.

11. Barros, R., Choi, W., Milewski, P.A. (2019) Strongly nonlinear effects on internal solitary waves in three-layer flows, *J. Fluid Mech.*, 883, A16-1–16–36.

This paper too will find its way into the discussion in a subsequent Chapter. I include it here to illustrate the so–called MCC theory, which extends the KdV theory discussed in this chapter for the case of layered fluids, and as such has strong proponents. I have always found its algebra heavy nature to be a price of entry that is too high for my tastes.

Chapter 3
Using Linear and Weakly Nonlinear Theory

Doing It Numerically Part I: Linear Theory

The essence of the linear theory (Boxes 3 and 4) is in splitting up the problem into manageable subproblems for the vertical structure and the horizontal structure. These smaller problems can be solved on their own and the solution then 'built up' from parts. For certain very special situations (e.g., the linear stratification, the two–layer stratification in Box 6) this can be done by hand, but for a general stratification numerical methods are necessary. Here we will focus on the eigenvalue problem for the vertical structure. We do this even though the general observation is that most of the observed internal wave dynamics in the coastal ocean or lakes is nonlinear, at least to some degree. The linear theory for the vertical structure is useful, because it gives us some general ideas of what the vertical structure of waves looks like. Moreover, the linear theory is used to compute the coefficients of the weakly nonlinear theories (Box 5) that capture at least some of the nonlinear wave physics. The challenge for numerical methods can be illustrated in two parts. First let us consider the case of no background current (Box 4), for which the governing eigenvalue problem reads (reproduced here for the reader's convenience) as

$$\phi_{zz} + \left(\frac{N^2(z)}{c_0^2} - k^2 \right) \phi = 0$$

where $\phi(0) = \phi(H) = 0$. Here we assume k^2 and $N^2(z)$ are given. It is possible to use interpolation to handle cases of $N^2(z)$ given from data, though $N^2(z)$ is not measured directly, and hence we leave the details of this process for another occasion. We thus assume $N^2(z)$ is known analytically. Though it turns out to be a bit of a pathological special case, the case of a constant $N^2(z) = N_0^2$ (or in other words a linearly stratified fluid) in the longwave limit is useful since it readily gives analytical solutions with easy interpretation, and we derived the solution in

M. Stastna, *Internal Waves in the Ocean*, Surveys and Tutorials in the Applied Mathematical Sciences 9, https://doi.org/10.1007/978-3-030-99210-1_3

the previous Chapter (2.5), and (2.6), which we reproduce here for the reader's convenience.

$$\phi_n(z) = \sin\left(\frac{n\pi}{H}z\right)$$

$$c_n = \frac{N_0 H}{n\pi}.$$

The above solution is useful for two reasons:

1. Solutions have more and more zeros in the interior of $[0, H]$ as n increases. Thus we can number the modes according to the number of interior zeros, plus one so that the first mode is labelled as $\phi^{(1)}$.
2. Higher modes have lower propagation speeds, or in other words mode-1 long waves are the fastest waves.

A linear stratification is not typical in the lab or the field, and so a natural goal is to solve numerically for a given $N^2(z)$. Matlab has several excellent built in functions for solving generalized eigenvalue problems for matrices that have the form

$$\mathbf{A}\vec{x} = \lambda\mathbf{B}\vec{x} \tag{3.1}$$

where boldface denotes a matrix. Thus we must convert an ordinary differential equation (henceforth ODE) eigenvalue problem into its matrix counterpart. We can rearrange the equation so that it looks a bit more like an eigenvalue problem,

$$c^2\left(\phi_{zz} - k^2\phi\right) = N^2(z)\phi.$$

Here the wave phase speed c is unknown, so $\lambda = c^2$ is the eigenvalue. The vector $\vec{x}$ can be thought of as the values of ϕ at our discretized points in $0 \le z_i \le H$. The remaining task is thus to discretize the second derivative operator. The Chebyshev pseudospectral methodology, [3], [12], is very effective for this problem, so following [3] let us call the differentiation matrix $\mathbf{D}$ and denote the second derivative by $\mathbf{D}^2$. We also create a diagonal matrix out of the values of $N^2(z_i)$ and denote the identity matrix as $\mathbf{I}$. Once this is done we have

$$\mathbf{B} = \mathbf{D}^2 - k^2\mathbf{I}$$

and

$$\mathbf{A} = \text{diag}\left[N^2(z_i)\right]$$

where $diag$ denotes the values of the function along the diagonal of a matrix with all off-diagonal entries equal to zero. Accounting for the boundary conditions is relatively easy once we note that the first and last entries of the vector $\vec{x}$ are the boundary points and hence are set to be zero by the boundary conditions. This means that the first and last rows of the matrices **A** and **B** are superfluous, and that the first and last column of these matrices are irrelevant since they always multiply 0. To put this into a more precise form, the case of homogeneous Dirichlet boundary conditions can be handled by simply removing the first and last rows and columns from the matrices, and the first and last entries of $\vec{x}$ from consideration.

The longwave limit, $k \to 0$, which is relevant for the weakly nonlinear, or KdV, theory is very simple to implement, but the necessary integrals used to compute the coefficients α and β do require a choice of integration rule to compute them. Thankfully, if one uses a pseudospectral Chebyshev method to discretize the derivatives, the corresponding integration method (i.e., Clenshaw–Curtis integration) uses the same grid and has the same spectral rate of convergence, meaning that excellent accuracy is achieved with even a moderate number of points.

For the first experiment, let us consider the effect of changing the wavelength, or in other words k. We will focus on first and second mode waves. The longwave limit is $k \to 0$, and accordingly the code starts with small wavelengths (large values of k) and gradually increases the wavelength. By running the script:

```
linear_wave_tutorial.m
```

you can see the resulting movie for yourself. The wave structure appears to be become more localized as k increases, and this can be confirmed in Fig. 3.1, which shows four sample values of wavelength.

We can get some analytical intuition for this based on the governing equation, which we rewrite as

$$\phi_{zz} + M\phi = 0.$$

If M were to be a constant, the sign of M would tell us whether we get oscillatory ($M > 0$) or exponential-type ($M < 0$) solutions. If k is fixed we thus see that the sign of

$$\left(\frac{N^2(z)}{c^2} - k^2 \right)$$

tells us whether we expect oscillatory or decaying solutions. In particular, for a single pycnocline stratification a useful model is

$$\bar{\rho}(z) = 1 + a \tanh\left(\frac{z - z_0}{d} \right)$$

where z_0 sets the pycnocline center, d sets its thickness, and a sets the dimensionless change in density across the pycnocline. The hyperbolic tangent function is defined in terms of exponentials and hence can be differentiated infinitely many times;

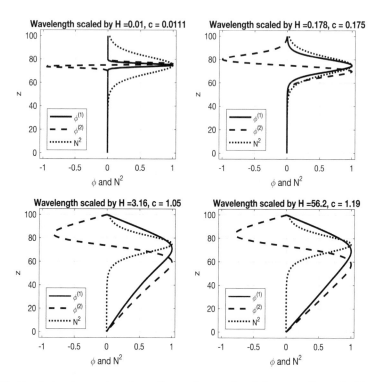

Fig. 3.1 The first and second mode profiles as the wavelength increases ($k \to 0$). The scaled stratification profile is indicated by a dotted line. Note how for shorter wavelengths the vertical profiles are more localized

sometimes the less than 100% mathematical term "smooth" is used. A direct calculation gives

$$N^2(z) = \frac{ag}{d}\operatorname{sech}^2\left(\frac{z - z_0}{d}\right).$$

This function has a maximum at $z = z_0$ and hence the switch from oscillatory to decaying solution occurs at a z value given by the transcendental equation

$$\frac{ag}{dc^2}\operatorname{sech}^2\left(\frac{z - z_0}{d}\right) = k^2.$$

The above results suggest the reader can explore at will. For example, the reader can readily modify the code so that $N^2(z) = N_0^2$ and derive a numerical version of the analytical results for the linearly stratified case.

Since a nearly two layer fluid is so typical in the laboratory it would be useful to both derive the limiting analytical solution (Box 6) and compare it to appropriate numerical results. To derive the analytical result note that the density profile for a two layer fluid may be written as

$$\bar{\rho}(z) = \rho_0 - \Delta\rho H(z - h_1)$$

where $H(z)$ is the Heaviside step function (admittedly a bit confusing given that H is the total depth) and h_1 is the lower layer thickness. Since $\bar{\rho}(z)$ is constant away from the interface, and we are in the longwave limit $k \to 0$, the governing ODE is very simple

$$\phi_{zz}^{(\text{two–layer})} = 0$$

and tells us that the solution is piecewise linear. To satisfy the boundary conditions we can write it as

$$\phi^{(\text{two–layer})}(z) = \begin{cases} \phi(h_1) + b(z - h_1) \text{ when } z > h_1 \\ az \text{ when } z < h_1. \end{cases}$$

Since we had set the maximum value of the eigenfunction to be one before, it makes sense to do it again, and indeed to set it to be equal to one at the interface. This also makes enforcing continuity very easy. A bit of algebra gives us

$$\phi^{(\text{two–layer})}(z) = \begin{cases} 1 - (z - h_1)/(H - h_1) \text{ when } z > h_1 \\ z/h_1 \text{ when } z < h_1 \end{cases} \tag{3.2}$$

and this can be readily confirmed to agree with the form in Box 6. One task remains, and that is to derive the wave propagation speed. To do so integrate the governing ODE

$$\int_0^H \left(\phi_{zz} + \frac{N^2(z)}{c_0^2}\phi \right) dz = 0$$

and notice that we can use the fact that the Dirac delta is the formal derivative of the Heaviside function, so that

$$\phi_z|_0^H = -\int_0^H (g\Delta\rho/c_0^2)\delta(z - h_1)\phi(z)dz.$$

The left hand side can be evaluated from the solution and the right hand side can be simplified using the property that

$$\int_0^H \delta(z - h_1)\phi(z)dz = \phi(h_1).$$

This gives

$$\frac{1}{h_2} + \frac{1}{h_1} = \frac{g\Delta\rho}{c_0^2}$$

and rearranging confirms the solution for the propagation speed in Box 6, namely

$$c_0 = \pm\sqrt{g\Delta\rho h_1 h_2 / H}. \tag{3.3}$$

A simple question we can use our numerical routines to check is "to what extent can a continuous stratification be considered to be well-approximated by the two layer one?" By running the script

```
twolayer_approx.m
```

we can answer this question. The live movie the script generates compares ϕ for the continuous case to ϕ for the two layer case as the interface thickness varies. Figure 3.2 shows four sample values of pycnocline thickness. The only real price to be paid is in computation time since in order to be careful about resolving the thin pycnocline, I run with 5 times as many points in the vertical. Nevertheless, on my laptop in under a minute the results readily pop up. What the results tell us is that if mode-1 waves are what we are interested in, then we can use either the two-

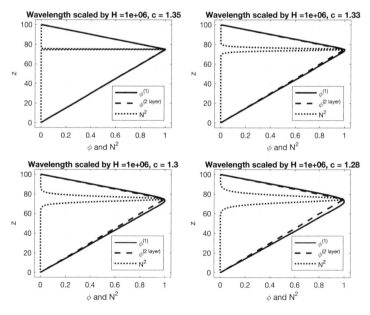

Fig. 3.2 The profile of the first mode profiles as the pycnocline thickness is changed. The scaled stratification profile is indicated by a dotted line and the two layer solution by a dashed line

layer theory or the numerical solution of the continuous approximation depending on which is more convenient.

The above examples are interesting in their own right, but they are practical as well. We can use them to extend the linear theory to describe field and laboratory observations. This is what we will discuss next. We will begin by considering the simplest case first, that being the KdV equation. Considering the KdV only, means α_2 can be ignored in boxes 6 and 7. Here is a collection of the relevant facts:

1. The two-layer theory is a good approximation of a single, smooth pycnocline.
2. The two-layer theory gives coefficients for the KdV theory in closed form.
3. The solitary wave solution of the KdV theory is in closed form.

This means we can get information about the solitary wave solutions of the KdV as a function of amplitude and layer thickness. The linear theory tells us that the linear propagation speed can be written as

$$c_0 = \sqrt{g \Delta \rho h_1 (H - h_1)/H}$$

and this clearly has a maximum when $h_1 = H/2$. Moreover c_0 exhibits no asymptotes or other pathological behavior. In contrast, rewriting the nonlinearity coefficient as

$$\alpha = \frac{3}{2} c_0 \frac{2h_1 - H}{h_1(H - h_1)}$$

shows that nonlinearity grows without bound as $h_1 \to 0^+$ or $h_1 \to H^-$. A more intuitive way to word this is to say that nonlinearity is very important when one of the layer depths is very small. The expression is also useful since it says that nonlinearity is not important, or α vanishes, when $h_1 = H/2$, or when the two layers are nearly the same thickness. Indeed the limit of nearly equal layers is what leads to the "improved" Gardner or mKdV equation, a fact we will revisit later.

The dispersive coefficient can be rewritten as

$$\beta = \frac{1}{6} c_0 h_1 (H - h_1)$$

and while there are no vertical asymptotes, the dispersive coefficient tends to zero when one of the two layers is very thin.

The above analysis is very satisfying as we have learned a lot without needing to solve anything complicated. The only issue is that the parameters often quoted (α and β) are not the actual physical quantities one would be interested in measuring. Better candidates for something to measure are the solitary wave solution profiles as a function of x, the horizontal coordinate, and the wave width and propagation speed. From Box 7 we see that the propagation speed is a linear function of amplitude, with an indirect dependence on the layer thickness through the expression for α. The dependence of γ, and the solitary wave profiles themselves, on the various

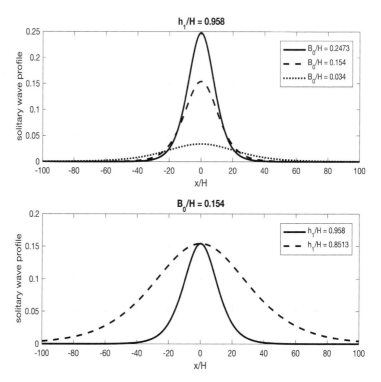

Fig. 3.3 Sample KdV solitary wave profiles as functions of amplitude (upper panel) and lower layer thickness (lower panel)

parameters is more complex, and best described from the graphs. In terms of the parameters in Box 7, the solitary wave width is given by $1/\gamma$.

By running the script:

```
twolayer_kdv_solwaves.m
```

you can see the results for yourself, or generate slightly different pictures (e.g., you could try creating 3D plots). Figure 3.3 shows sample KdV solitary wave profiles as functions of amplitude (upper panel) and lower layer thickness (lower panel). Increasing amplitude with a fixed layer thickness leads to taller and narrower waves. This is commonly presented in mathematical sources on the KdV equation, which treat a version of the KdV without the physical parameters (what mathematicians call "standard form"). The lower panel is more interesting. It shows that at a fixed amplitude the wave width is quite sensitive to the lower layer thickness. In practice this would mean that solitary waves with very different widths could be considered during different seasons, as the stratification changes.

A different look at the variations in wave widths is presented in Fig. 3.4. Here the upper panel again reflects the fact that taller KdV solitary waves are narrower.

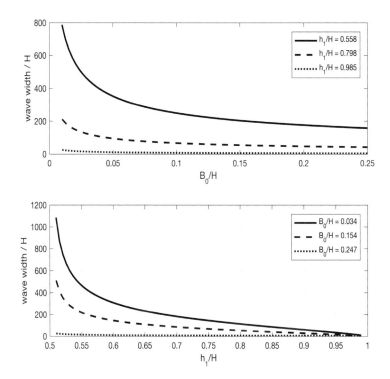

Fig. 3.4 Sample KdV solitary wave width profiles as functions of amplitude (upper panel) and lower layer thickness (lower panel)

Moreover, one can see just how profound the broadening is as h_1 approaches half of the total depth (or in other words the two layers are nearly the same depth).

What I would like to do next is to move on to the problem with a background current. The relevant eigenvalue problem is shown in Box 3, and is called the Taylor–Goldstein (T-G) equation. You can think of it is a natural extension of the case without a background current, shown in Box 4, or as my ordering suggests a more general case from which the case of no background current can be extracted by setting $U(z) = 0$. The necessary algebra to derive the T-G equation is, however, more complicated, and numerically, some inventiveness is required as well. To see this recall the form of the T-G equation (derivation culminating in Eq. (2.4) was presented in the previous Chapter),

$$\phi_{zz} + \left(\frac{N^2(z)}{(c-U)^2} + \frac{U''}{c-U} - k^2 \right) \phi = 0$$

and notice that in the denominator we have a term of the form $(c - U(z))^2$. This means that were we to use our Chebyshev differentiation matrices for the derivatives, and rearrange the problem to look like a standard linear algebra problem we would get a polynomial eigenvalue problem

$$\left(c^2 A_2 + c A_1 + A_0\right) \phi = 0$$

for some expressions A_2, A_1, A_0 (you can work them out as an exercise). Polynomial eigenvalue problems are generally not taught in standard linear algebra courses, though they are well-studied in the linear algebra community, with some algorithms available for their solution (type *help polyeig* in Matlab to see how Matlab does it). The more standard numerical approach is to convert the quadratic eigenvalue problem into a generalized eigenvalue problem. I like this approach because it points out just how strange equations like the T-G are. Basically the T-G is an attempt to convert as much of a problem as possible (traditionally the stability of a stratified shear flow) into a second order ODE. Prior to the advent of computation this made a lot of sense, as the theory of second order ODEs was very well developed. However, with computation being readily available we do not need to feel shackled by the same constraints. So let us revisit the derivation of the T-G equation.

Assuming infinitesimal perturbations to a background velocity state $(U(z), 0)$ and a background density stratification $\bar{\rho}(z)$ led to the linearized stratified Euler equations. Then assuming perturbations of "plane wave" type we were able to derive two equations for the vertical structure of the streamfunction and buoyancy (2.3), which are reproduced here for the reader's convenience as

$$-c\hat{d} + \hat{d}U + \phi \frac{N^2}{g} = 0,$$

$$-c(\phi_{zz} - k^2 \phi) + U(\phi_{zz} - k^2 \phi) - U_{zz}\phi = \hat{d}g.$$

Recall, that these two equations can be further rearranged and combined together, in a process that yields the T-G equation. What we want to pursue instead is writing the system of equations as a generalized eigenvalue problem, but this time in block form. As in the above work on the case without a background current, let $\mathbf{D}^2$ be the second order Chebyshev differentiation matrix. Once again, the boundary conditions are incorporated using the simple trick of removing the first and last columns and rows. We denote the identity matrix by $\mathbf{I}$, and in fact, we can make our life easier by defining a reusable piece

$$\mathcal{L} = D^2 - k^2 \mathcal{I}. \tag{3.5}$$

Next suppose the variables $\phi(z)$ and $d(z)$ are evaluated at the Chebyshev points and arranged in a single vector one after the other. We then find a single eigenvalue problem, whose matrix form is given by the standard form

$$\mathcal{A}\vec{v} = c\mathcal{B}\vec{v}$$

with the matrices and vector given in block form as

$$\mathcal{A} = \begin{pmatrix} diag(N^2(z)) & diag(U) \\ diag(U)\mathcal{L} - diag(U_{zz}) & -\mathcal{I} \end{pmatrix}, \tag{3.6}$$

$$\mathcal{B} = \begin{pmatrix} \mathbf{0} & \mathcal{I} \\ \mathcal{L} & -\mathbf{0} \end{pmatrix} \tag{3.7}$$

and

$$\vec{v} = \begin{pmatrix} \phi(z) \\ g\hat{d}(z) \end{pmatrix}. \tag{3.8}$$

Here is the Matlab code snippet for the process

```
% make up the matrices for the e-val prog.
% eye(N) makes an N by N identity
% diag(vector) makes a diagonal matrix with the vector along the diagonal
p1=D2-k2*eye(N-1);
a11=diag(u)*p1-diag(uzz);
a12=-eye(N-1);
a21=diag(nn);
a22=diag(u);
b11=p1;
b22=eye(N-1);
% define A
A=zeros(2*(N-1),2*(N-1));
A(1:N-1,1:N-1)=a11;
A(1:N-1,N:2*(N-1))=a12;
A(N:2*(N-1),1:N-1)=a21;
A(N:2*(N-1),N:2*(N-1))=a22;
% define B
B=zeros(2*(N-1),2*(N-1));
B(1:N-1,1:N-1)=b11;
B(N:2*(N-1),N:2*(N-1))=b22;
% Solve the e-val prob
  [Xim eedim]=eig(A,B);
  % Redimensionalize and post-process
  ee=diag(eedim)*bigu;
[sorteepos sortindpos]=sort(real(ee),'descend');
clpos=ee(sortindpos(1));
% This is the line that pulls out the phi(z) bit from the eigenvector
philpos=[0; Xim(1:N-1,sortindpos(1)); 0];
Elpos=clpos*philpos./(clpos-uphysical);
```

The numerical experiment we wish to carry out with the new solver considers a fixed stratification and allows the user to change the strength of a linear background current. The script is called:

```
spectraltg_tutorial.m
```

One of the challenges with presenting the results with a background shear is the fact that the same eigenvalue problem applies for both waves and instabilities (which grow in time). This book does not get into the details of hydrodynamic stability and so care must be taken to make sure that the results of the computation are not taken as completely general. For strong enough shear one must consider complex eigenvalues and eigenfunctions. In practice, the readers must determine for themselves whether a solution makes sense, and this is an essential part of learning a given topic. In particular, if there is some z^* for which $U(z^*) = c$ the T-G equation fails due to the $U(z) - c$ terms in the denominator. For a given background current this typically means that higher modes, which have smaller c, are more influenced by a particular background current. The reader is encouraged to slowly increase the parameter $uamp$ in the code. Note in particular the way that a higher $uamp$ yields mode-1 eigenfunctions whose maximum is displaced from the pycnocline mid-depth. Moreover at these values the code fails to find one of the two mode-2 eigenvalue/eigenfunction combinations.

One way to analyze the results in the script is to consider the sign of the shear. Since the background current is very simple, the shear is a constant. The sign of the shear is set by the sign of $uamp$. Figures 3.5 and 3.6 show how a moderate shear modifies the mode-1 and mode-2 eigenfunction profiles, respectively. The propagation speeds are given in the right panel titles. Mode-1 propagation speeds are

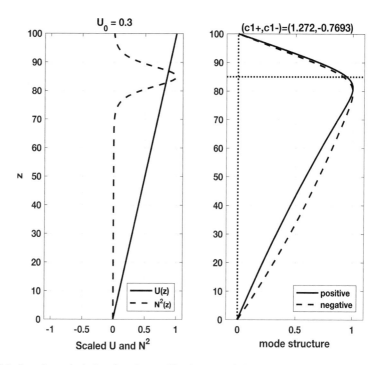

Fig. 3.5 Sample mode-1 eigenfunction profiles ($uamp = 0.3$)

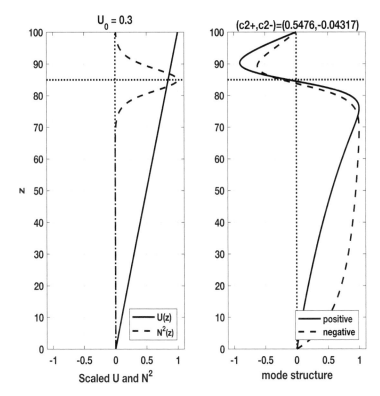

Fig. 3.6 Sample mode-2 eigenfunction profiles ($uamp$ = 0.3). Note how close to zero the propagation speed of the leftward propagating wave is

visibly modified, but mode-2 propagation speeds are modified to the point where the leftward propagating wave has a propagation speed that is nearly zero. This suggests that this mode is close to breaking down if the shear is increased even slightly (the reader could try using the code to figure out exactly at what value of $uamp$ the solver fails to find the mode-2 wave). If the mode-1 wave is all we care about, then we can in fact go to much stronger background currents. Figure 3.7 shows the result for the mode-1 waves when $uamp$ = 0.8. The vertical structure of leftward and rightward propagating waves is now fundamentally different.

Shear makes the problem quite complicated rather quickly. We will revisit the role of shear issue when we discuss the solutions of the DJL equation, and in the third of the open ended mini-projects below.

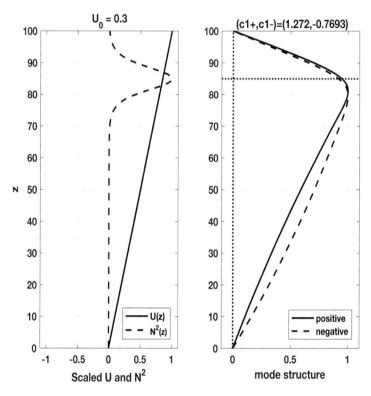

Fig. 3.7 Sample mode-1 eigenfunction profiles ($uamp = 0.8$). There are no mode-2 or higher mode waves for this case

Mini-Project 1: WNL Coefficients Versus Pycnocline Depth:
The first mini-project considers the variation of the coefficients of the KdV theory as the pycnocline center is moved. The pycnocline thickness remains fixed. The script is called:

```
wnl_coefs_vs_depth.m
```

and when you run it the eigenvalue problem is repeatedly solved for various depths of the pycnocline center. The coefficients of the KdV are solved for at each height. This involves a numerical integration, which is carried out using the so-called Clenshaw–Curtis nodes and weights. This is the pseudospectral counterpart to the Chebyshev differentiation scheme that we use to convert the derivative operator to a matrix. The script plots the coefficients and propagation speeds at each height and scales by the results at a particular depth. This depth is indicated with a horizontal blue line. For each of the two modes shown (mode-1 and mode-2) you should ask yourself some simple questions and see if the plot allows you to answer them for yourself:

1. At what height is the propagation speed maximum?
2. At what height of pycnocline center does the nonlinear coefficient have a maximum magnitude?
3. What, if anything happens when the pycnocline is centered at the mid-depth?

Mini-Project 2: WNL Coefficients Versus Pycnocline Thickness:
The second mini-project considers the variation of the coefficients of the KdV theory as the pycnocline width is increased. The script is called:

```
wnl_coefs_vs_thickness.m
```

and when you run it the eigenvalue problem is repeatedly solved for a fixed depth of the pycnocline center, but varying pycnocline thickness. There are some interesting differences to note between how mode-1 and mode-2 waves behave as the pycnocline thickness changes, which I leave for the reader to puzzle out. For mode-1 waves, the noteworthy feature I wish to point the reader to is the decrease in the nonlinearity coefficient as the thickness increases.

Mini-Project 3: Near Surface Jets in the Taylor–Goldstein Equation:
The third mini-project considers a more realistic $U(z)$ in the T-G equation.
The script is called:

```
spectraltg_mini_project.m
```

and when you run it the eigenvalue problem is solved for $U(z)$ that is either
near surface or near bottom intensified. You can try the various combinations
of background current amplitude and shear layer thickness that are left as
comments for you. Then identify a question (e.g., for a fixed d_j, what is the
largest *uamp* for which you can compute mode-2 waves?) and run the script
repeatedly to answer your question. Use appropriate plots to document what
you found.

Doing It Numerically Part II: WNL

Now that we have achieved some level of understanding of the vertical structure
of the modes and how the coefficients of the weakly nonlinear theory change as the
details of the stratification change, we are in a position to actually look at how waves
evolve. While we could start with linear theory, we choose to begin right away with
a consideration of the effects of finite amplitude and finite wavelength, or in other
words the effects of nonlinearity and dispersion.

In this section we will consider the BBM equation (shown in Box 5 and originally
derived in [13]) reproduced here for the reader's convenience,

$$B_t + c_0 B_x + \alpha B B_x - \frac{\beta}{c_0} B_{xxt} = 0$$

using the coefficients of two layer theory (Box 6). We will again use spectral
methods, though of a slightly more traditional sort. As two basic points of notation
let us denote the wavenumber as k, and the spatial Fourier transform of a variable
by an overbar. This means

$$\bar{u}(k, t) = \int_{\infty}^{\infty} \exp(-ikx) u(x, t) dx,$$

with the FFT providing a numerical implementation . There is some pretty theory
behind all this, and Trefethen does a great job presenting it in [3], but for the present
we will really just need the fact that the Fourier transform converts derivatives to
multiplication by ik, or

$$\overline{u_x} = ik\bar{u}.$$

We note one minor point of algebraic trickery, namely that

$$\overline{BB_x} = \frac{1}{2}\overline{(B^2)_x} = \frac{ik}{2}\overline{B^2}.$$

With this in hand we Fourier transform the BBM equation and rearrange to find

$$\left(1 + \frac{\beta}{c_0}k^2\right)\overline{B_t} = -c_0ik\bar{B} - \frac{1}{2}\alpha ik\overline{B^2}.$$

It is a simple matter to isolate for the time derivative so that

$$\overline{B_t} = \frac{1}{\left(1 + \frac{\beta}{c_0}k^2\right)}\left[-c_0ik\bar{B} - \frac{1}{2}\alpha ik\overline{B^2}\right]. \tag{3.9}$$

Recall that β is always positive and hence the denominator of the expression never vanishes. With this Fourier transformed equation we can note a couple of things. First, that we can very easily construct an explicit method for its numerical approximation. If we discretize the time derivative by a method of our choice, and evaluate the right hand side at a time $t = t_n$ using the FFT we should be able to code up the time integration without any trouble. Second the factor

$$\frac{1}{\left(1 + \frac{\beta}{c}k^2\right)}$$

on the right hand side means that shorter wave components, or in other words those with a larger k^2, are multiplied by a smaller and smaller number. This is very convenient since any finite grid has a cutoff k above which no information is known, and because the nonlinear term sends information to higher and higher wavenumbers as time goes on (this is called aliasing in the numerics literature). Indeed most numerical instabilities start their short, nasty life at the grid scale and while the above is not a rigorous proof, it is certainly suggestive that we expect the method to work.

Before we turn to a discussion of implementation, let us contrast what we found for the BBM equation with what happens for the KdV equation,

$$B_t + c_0B_x + \alpha BB_x + \beta B_{xxx} = 0.$$

Fourier transforming gives

$$\overline{B_t} = -ik(c_0 - k^2\beta)\bar{B} - \frac{\alpha}{2}ik\overline{B^2}.$$

If we ignore the nonlinear term for the moment, we can write the equation in a suggestive way,

$$\overline{B_t} = -ik(c_0 - k^2\beta)\bar{B} = -ikc^{(dispersive)}(k)\bar{B},$$

where $c^{(dispersive)}(k)$ gives the phase speed of a particular wave length. However, a brief examination of the two layer case in Box 6 reveals that $beta > 0$, meaning that there is some k above that $c^{(dispersive)}(k) < 0$. This is not very satisfactory on physical grounds, since all the WNL theories assumed waves are propagating to the right. Moreover, $c^{(dispersive)}(k)$ keeps decreasing (i.e., moves faster and faster to the left) as k increases and hence the shortest scales (largest k) put the strongest restrictions on time step (recall that one way to state the CFL condition says that a wave cannot move through a grid box in one time step). This is why the KdV is often said to be challenging numerically, and why the BBM equation is preferable.

OK, let us turn to the implementation. Here is the Matlab code snippet for the evolution using explicit Euler for a first try (generally not recommended for production runs):

```
bbmfact =1./(1+( betatwolayer / ctwolayer ).* ks2 );
% start with Euler time stepping
for  ii =1:numouts
    for  jj =1:numstps
      t=t+dt ;
      % explicit method for BBM
      B1lin=-ctwolayer*sqrt(-1)*ks.* fft (B1 );
      B1nl=-0.5*alphatwolayer*sqrt(-1)*ks.* fft (B1.^2);
      B1 = B1+dt*real ( ifft ( bbmfact.*( B1lin+B1nl )));
    end
end
```

The first thing we are going to show are two basic evolution cases. The script is called:

bbmpolarity.m

and will produce a movie when you run it. We consider a two layer stratification in water 100 m deep with an upper layer that is 20 m thick. We consider two initializations that vary only in terms of their polarity. The negative polarity is the "right" polarity for solitary waves to evolve and indeed three rank ordered solitary waves are clearly evident in solid in the movie. The positive polarity clearly shows that the steepening is toward the back of the wave. The dispersive effects do not really visibly kick in until the wave is quite steep and at this point a wave train is

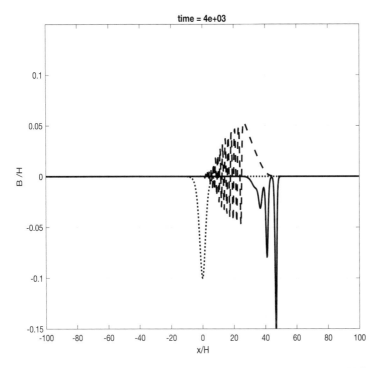

Fig. 3.8 The evolution of the BBM equation for two different polarities. The solid line is the control case, and the dashed line is the positive polarity case. The dotted line is a reminder of the initial conditions for the control case

evident. The shape of the wave train is quite particular in that a long, slow slope forms the leading edge of the wave, while the back end is an actual train of finite wavelength waves. The term often used for this in the literature is "undular bore" (Fig. 3.8).

The second example considers the effect of initial condition amplitude. The script is called:

```
bbmamplitude.m
```

The live movie shows three panels. The convention is the same as in the previous example, with the solid line being the "control" and the dashed line being the "manipulation," which in this case means an initial amplitude reduced by 50%. The upper panel shows all curves, the middle panel the control case, and the lower panel the manipulation. You can see that the leading solitary wave in the smaller amplitude case more or less matches the second solitary wave in the control case. You can change the amplitude or even the polarity of the initial condition to see how the nonsolitary wave case evolves. You will want to change the axis command as well. Two figures from this case are shown below, but for these I only show the middle (the control case) and lower (the reduced initial amplitude case) panels from the movie.

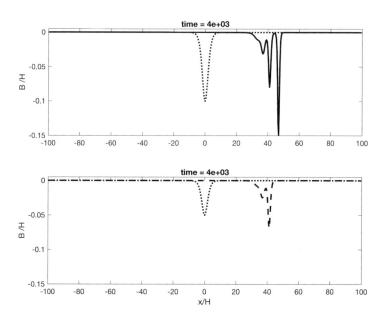

Fig. 3.9 The evolution of the BBM equation for a change in initial condition amplitude. For both cases the initial condition is shown by the dotted line

Figures 3.9 and 3.10 show the development of the waves at two different times. Interestingly at the later output time, the leading wave from the smaller amplitude case is very similar to the second solitary wave in the wave train that emerges from the control case. The nonlinear nature of the governing equation is reflected in the fact that the amplitude of the largest solitary wave is larger than the amplitude of the initial conditions.

The third example considers the influence of width of the initial conditions in setting the size and number of solitary waves that develop. The script is called:

```
bbmwidth.m
```

I modify the initial conditions a little bit, choosing a smooth approximation to a "square wave" initial condition. The "square wave" is commonly used in inverse scattering theory since it allows for a quasi-analytical solution (albeit after a lot of algebra). The primary difference between the wider and narrower initial condition is in the number and amplitude of the solitary waves that emerge from the initial condition. There is also a small amplitude dispersive wave train visible well behind the rank ordered train of solitary waves.

Figures 3.11 and 3.12 show the development of the waves at two different times. The emergence of the initial waves is slower for the wider initial condition. At the later time shown, not only have more solitary waves emerged in the wider case (6 as opposed to 3), but the largest wave is larger, even though the amplitude of the initial condition is the same.

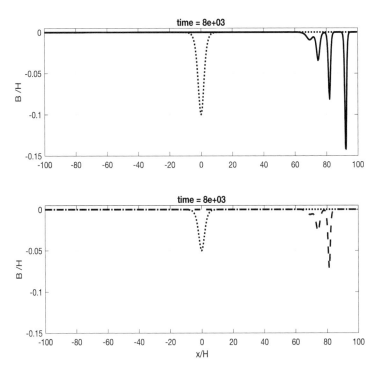

Fig. 3.10 The evolution of the BBM equation for a change in initial condition amplitude. For both cases the initial condition is shown by the dotted line. At this later time the solitary waves are well developed

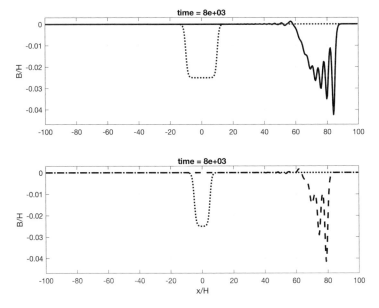

Fig. 3.11 The evolution of the BBM equation for a change in initial condition width. For both cases the initial condition is shown by the dotted line

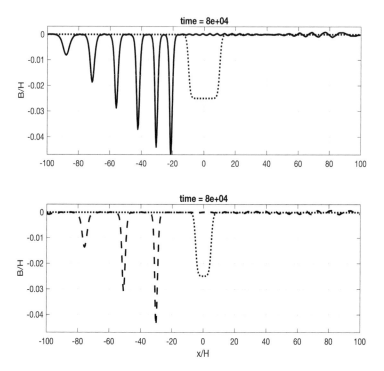

Fig. 3.12 The evolution of the BBM equation for a change in initial condition width. For both cases the initial condition is shown by the dotted line. At this later time the solitary waves are well developed

Mini-Project 1: Solitary Wave Collisions The mini-project for this section involves the collision of a large solitary wave with several copies of a small solitary wave. The script is called:

```
bbmcrash.m
```

For linear waves we would expect waves to just pass through each other with no effect at all. But there is no reason at all why this should be the case for nonlinear wave. Yet it is precisely the point of soliton theory that this does occur. Mathematically, this is the whole point of the inverse scattering formalism and other theories for the KdV (and similar) equations. But the real power, in the author's opinion anyway, is that it very nearly holds without all that attendant mathematics. The BBM equation does not have an inverse scattering formalism, so how will it do when it comes to collisions? We start the simulation with one large solitary wave and, for the perturbation case, three identical small solitary waves found to the right of the large one. The large wave propagates through the small waves. For most of the movie you can only see the large wave in red, meaning the case without any small waves is no different from the one with the small waves. However, after all three interactions you can see a small difference between the blue and red wave. If you focus on using the Matlab magnify tool you can see, somewhat amazingly that the blue wave is actually ahead of the red wave. This means the collision has made the large wave end up ahead of where it would be were it to propagate all by itself. You can also see that what the lack of an inverse scattering theory means in practice, since the blue wave is reduced in amplitude. The reduction is small, but it is not zero.

Concluding Thoughts

The primary purpose of this chapter has been to demonstrate just how easily well-written code can give results. Indeed the results come so quickly that for the time dependent cases live movies are possible even on a laptop. This means that scientific hypotheses can be readily formed, and the overhead for computing both a 'control' and a 'manipulation' is inconsequential. In terms of ocean science the various case studies allow for various detailed conclusions, but the four broader conclusions are

1. The various modes behave quite differently from one another, with mode-1 being exceptional.
2. The nonlinear/dispersive nature of the ocean waveguide means that trains of solitary waves are the rule, as opposed to an interesting exception.

3. Nonlinear waves can be larger in amplitude than the initial state from which they form.
4. The rank ordered train of solitary waves is a general feature observed for many different initial conditions.

These conclusions are powerful, and quite different from a linear theory point of view. They do however involve mathematical approximations that are opaque. The key aspect of these approximations to note here is that the KdV, Gardner, BBM, or other model equations DO NOT come with an error bound (e.g., in water 100 m deep use only for waves with an amplitude of 10 m or less). Moreover, given the ubiquity of the rank ordered train of "bell-shaped" solitary waves, it should be no surprise that something like KdV solitons are measured in the ocean. In the next chapter we will explore an alternative theory that does not require an upper bound in amplitude, and discuss some of the ways that measurements could go beyond the KdV soliton in terms of the description of real waves.

Literature 3

12. J. P. Boyd. (2001) Cheyshev and Fourier spectral methods. Dover publications.
Boyd's book is the Yang to Trefethen's [3] Ying. It is broad in coverage, and agnostic as to implementation language. As such, it makes for interesting re-reading many years after you first acquire a copy.
13. T. B. Benjamin, J. L. Bona, and J. J. Mahony. (1972) Model equations for long waves in nonlinear dispersive systems. Phil. Trans. Roy. Soc. A, 227:47–78.
This paper derived the eponymous BBM equation that regularizes (i.e., makes the linear dispersion behavior more in line with what is observed) the KdV equation.

Chapter 4
Exact Internal Solitary Waves

DJL Theory

Dubreil–Jacotin Long, or DJL, theory makes a very different trade-off compared to the weakly nonlinear theory presented in the previous chapters. No longer do we seek equations valid only in certain asymptotic limits (small amplitude, long wave). Instead, we want to a priori choose a solution type, namely a translating wave of permanent form (i.e., a solitary wave), and derive the equation that governs it. Thus the DJL equation is equivalent to the full set of stratified Euler equations, albeit for this special class of solutions.

The DJL equation is immensely practical because it has a tidy numerical solution technique. Its derivation is algebraically challenging and rarely presented. It is for this reason that I include all the details below. I do try to discuss some of the reasons for the various tricks as they are employed. The first of these is that even though we may not know the wave speed ahead of time, if we switch into a frame moving with this wave speed, the flow field is steady.

We consider a non-rotating, incompressible, inviscid fluid under the Boussinesq approximation (see Box 1) in a fixed frame of reference with the origin at the ocean floor. The x-axis is taken as parallel to the flat ocean bottom with the z-axis pointing upward. The governing equations are repeated here for the reader's convenience:

$$u_t + uu_x + wu_z = -\frac{p_x}{\rho_0} \tag{4.1}$$

$$w_t + uw_x + ww_z = -\frac{p_z}{\rho_0} - \frac{\rho g}{\rho_0} \tag{4.2}$$

$$\rho_t + u\rho_x + w\rho_z = 0 \tag{4.3}$$

$$u_x + w_z = 0 \tag{4.4}$$

© The Author(s), under exclusive license to Springer Nature Switzerland AG 2022
M. Stastna, *Internal Waves in the Ocean*, Surveys and Tutorials in the Applied
Mathematical Sciences 9, https://doi.org/10.1007/978-3-030-99210-1_4

where u and w are the horizontal and vertical velocities, respectively, p is the pressure, ρ is the total density, ρ_0 is a constant reference density, g is the acceleration due to gravity, and subscripts denote partial derivatives. With a horizontal background current, we decompose the velocities as

$$u(x, z, t) = V(z) + \tilde{u}(x, z, t)$$

$$w(x, z, t) = \tilde{w}(x, z, t)$$

where $V(z)$ represents the sum of background shear current $U(z)$ and the shift into the frame moving with the wave, $-c$, while $\tilde{u}(x, z, t)$, $\tilde{w}(x, z, t)$ are the wave-induced parts of the flow. Note that $V'(z) = U'(z)$.

From the divergence-free flow condition, we can define a streamfunction $(\tilde{u}, \tilde{w}) = (\psi_z, -\psi_x)$ for the wave-induced flow. In addition, we introduce a background streamfunction $\psi^b = \int V(z)dz$ so that $\psi_z^b(z) = V(z)$. The total streamfunction is then given by

$$\psi^T(x, z) = \psi^b(z) + \psi(x, z),$$

and the steady state (recall we mean "steady" in a frame moving with the wave) version of the density equation reads

$$J\left(\rho, \psi^T\right) = 0,$$

where $J(a, b) = a_x b_z - a_z b_x$ is the Jacobian operator. If $J(a, b) = 0$, then the two functions are dependent, or $a = F(b)$ for some hitherto unknown function $F(.)$.

We write the density in terms of the far upstream profile,

$$\rho(x, z) = \rho_0 \bar{\rho}(z - \eta), \tag{4.5}$$

where $\eta(x, z)$ denotes the isopycnal displacement from the upstream rest height and the ρ_0 is the dimensional reference density value. For our isolated waves, we assume that $\eta, \eta_x, \eta_z \to 0$ as $|x| \to \infty$. Now we use the way we rewrote the density and some simple algebra to find that

$$J\left(\rho, \psi^T\right) = \rho_0 \bar{\rho}'(z - \eta)J\left(z - \eta, \psi^T\right) = 0.$$

Since $\bar{\rho}'(z - \eta) \neq 0$ everywhere, $J\left(z - \eta, \psi^T\right) = 0$, and thus

$$\psi^T(x, z) = F(z - \eta)$$

for some function $F(.)$. Next note that far upstream, or as $x \to -\infty$, we have
$\eta \to 0$, $(u, w) \to (V(z), 0)$ and $\psi^T \to \psi^b(z)$. Hence,

$$F(z) = \psi^b(z)$$

and

$$F(z - \eta) = \psi^b(z - \eta)$$

so that

$$\psi^T(x, z) = \psi^b(z - \eta). \tag{4.6}$$

This relation can be used to compute the velocity field via the chain rule, once
the isopycnal displacement η is known. It is worth going over the nature of the trick
since variants of it will be used again below. The essence is to find relations between
groupings of quantities whose Jacobian is zero. This means they are functionally
dependent. To determine the form of the function, we move to the only place where
we know something about our solution, namely far enough upstream where there
is no wave and the flow field is given by the background state and the shift to the
frame moving with the wave speed.

Next, we turn to the momentum equations. Substituting the streamfunction,
ignoring time derivatives, and taking the curl, we find that

$$(\psi_{xx} + \psi_{zz})_x \, \psi_z^T - \left(\psi_{xx} + \psi_{zz}^T\right)_z \psi_x = \frac{\rho_x g}{\rho_0},$$

which is equivalent to (deriving this would be a useful exercise)

$$J\left(\nabla^2 \psi^T, \psi^T\right) + J\left(\frac{gz}{\rho_0}, \rho\right) = 0. \tag{4.7}$$

Let $\sigma^T = \nabla^2 \psi^T = \nabla^2 \psi^b(z - \eta)$ denote the total vorticity (note that the frame shift
has no vorticity change associated with it), and note that

$$\sigma^T = V_z(z - \eta)\left(\eta_x^2 + (1 - \eta_z)^2\right) - V(z - \eta)\nabla^2\eta. \tag{4.8}$$

Now, write (4.7) as a vorticity equation,

$$J\left(\sigma^T, \psi^T\right) + J\left(\frac{gz}{\rho_0}, \rho\right) = 0, \tag{4.9}$$

and work with the $J\left(\frac{gz}{\rho_0}, \rho\right)$ term:

$$J\left(\frac{gz}{\rho_0}, \rho\right) = J\left(\frac{gz}{\rho_0}, \rho_0\bar{\rho}(z-\eta)\right)$$

$$= \rho_0\bar{\rho}'(z-\eta)J\left(\frac{gz}{\rho_0}, z-\eta\right)$$

$$= \rho_0\bar{\rho}'(z-\eta)J\left(\frac{g}{\rho_0}(z-\eta+\eta), z-\eta\right)$$

$$= \rho_0\bar{\rho}'(z-\eta)\left[J\left(\frac{g}{\rho_0}(z-\eta), z-\eta\right) + J\left(\frac{g}{\rho_0}\eta, z-\eta\right)\right]$$

$$= \bar{\rho}'(z-\eta)J\left(g\eta, z-\eta\right).$$

At this point, we want to bring the $\bar{\rho}'(z-\eta)$ inside the Jacobian. Notice that

$$J\left(g\bar{\rho}'(z-\eta)\eta, z-\eta\right) =$$

$$\left[g\bar{\rho}''(z-\eta)(-\eta_x)\eta + \frac{g}{\rho_0}\bar{\rho}'(z-\eta)\eta_x\right](1-\eta_z)$$

$$- \left[g\bar{\rho}''(z-\eta)(1-\eta_z)\eta + g\bar{\rho}'(z-\eta)(1-\eta_z)\right](-\eta_x)$$

$$= g\bar{\rho}'(z-\eta)\eta_x(1-\eta_z+\eta_z)$$

$$= g\bar{\rho}'(z-\eta)J\left(\eta, z-\eta\right),$$

so that

$$J\left(\frac{gz}{\rho_0}, \rho\right) = J\left(\frac{g}{\rho_0}\bar{\rho}'(z-\eta)\eta, z-\eta\right).$$

Next, work with $J\left(\sigma^T, \psi^T\right)$,

$$J\left(\sigma^T, \psi^T\right) = J\left(\sigma^T, \psi^b(z-\eta)\right)$$

$$= V(z-\eta)J\left(\sigma^T, z-\eta\right)$$

Again, we want the $V(z-\eta)$ term inside the Jacobian so notice that

$$J\left(V(z-\eta)\sigma^T, z-\eta\right) = \left[V_z(z-\eta)(-\eta_x)\sigma^T + V(z-\eta)\sigma_x^T\right](1-\eta_z)$$

$$- \left[V_z(z-\eta)(1-\eta_z)\sigma^T + V(z-\eta)\sigma_z^T\right](-\eta_x)$$

$$= V(z - \eta) \left[\sigma_x^T (1 - \eta_z) - \sigma_z^T (-\eta_x) \right]$$

$$= V(z - \eta) J \left(\sigma^T, z - \eta \right).$$

We now have

$$J \left(\sigma^T, \psi^T \right) = J \left(V(z - \eta) \sigma^T, z - \eta \right),$$

and after the appropriate substitutions, (4.9) becomes

$$J \left(V(z - \eta) \sigma^T + g \bar{\rho}' (z - \eta) \eta, z - \eta \right) = 0$$

or

$$V(z - \eta) \sigma^T + g \bar{\rho}'(z - \eta) \eta = G(z - \eta).$$

As $x \to -\infty$, we know that $\eta \to 0, \eta_x \to 0, \eta_z \to 0, \sigma^T \to V_z(z)$. So $G(z) = V_z(z) V(z)$ and $G(z - \eta) = V_z(z - \eta) V(z - \eta)$. Hence,

$$V(z - \eta) \sigma^T + g \bar{\rho}'(z - \eta) \eta = V_z(z - \eta) V(z - \eta).$$

After substituting (4.8) for σ^T, rearranging, and substituting the squared buoyancy frequency $N^2(z) = -g \bar{\rho}'(z)$, we have

$$\nabla^2 \eta + \frac{V_z(z - \eta)}{V(z - \eta)} \left[1 - \left(\eta_x^2 + (1 - \eta_z)^2 \right) \right] + \frac{N^2(z - \eta)}{V(z - \eta)} \eta = 0$$

or in terms of the unknown c and the background current profile $U(z)$,

$$\nabla^2 \eta + \frac{U_z(z - \eta)}{(U(z - \eta) - c)} \left[1 - \left(\eta_x^2 + (1 - \eta_z)^2 \right) \right] + \frac{N^2(z - \eta)}{(U(z - \eta) - c)^2} \eta = 0$$
$$(4.10)$$

Here the upstream current contains the unknown wave speed c. The DJL equation can be simplified, but we will not bother since it is much cleaner to do this in the case of no background shear current.

In the case of no far upstream current, $V(z) = -c$, where c is the wave speed and the above derivation simplifies considerably. The density equation yields the relation

$$\psi = -c(z - \eta).$$

between the streamfunction and the isopycnal displacement, while the vorticity equation no longer involves evaluation of the upstream velocity profile at its upstream height. This means that the amount of algebra is greatly reduced, yielding the more familiar form of the DJL equation

$$\nabla^2 \eta + \frac{N^2(z - \eta)}{c^2} \eta = 0. \tag{4.11}$$

The DJL equation was first derived in [14], though the equation's popularization in the English speaking world did not come until after the work of Long [15] who wrote extensively about the version of the DJL equation with N^2 constant. This version of the equation is not particularly useful in the oceans, though it remains popular in certain branches of the meteorological literature.

This equation has an interesting symmetry that relates waves in a situation with the pycnocline below the mid-depth to waves in a situation with the pycnocline above the mid-depth. This can be demonstrated as follows. Let $y = H - z$, and consider

$$e(x, z) = -\eta(x, y) \tag{4.12}$$

where η is a solution of (4.11) with a particular $N^2(z)$ profile. Next consider a second stratification (a nice exercise is confirming that such a stratification is still stable) defined by the relation

$$\tilde{N}^2(z) = N^2(y).$$

Applying the chain rule, we find that

$$\nabla^2 e = e_{xx} + e_{zz} = -\eta_{xx} - \eta_{yy}.$$

Moreover,

$$\tilde{N}^2(z - e) = N^2(H - (z - e)) = N^2(y + e) = N^2(y - \eta(x, y)).$$

Thus

$$\nabla^2 e + \frac{\tilde{N}^2(z - e)}{c^2} e = -\eta_{xx} - \eta_{yy} - \frac{N^2(y - \eta)}{c^2} \eta(x, y) = 0$$

where the last step just used (4.11) with y as the vertical variable. The relation (4.12) means that if we have a wave of one polarity for a single pycnocline stratification centered at $z = z_0$ (e.g., above the mid-depth, $z_0 > H/2$), we can immediately (i.e., without solving the DJL equation again) find waves with the **opposite** polarity with the exact same shape for a stratification centered at $z = H - z_0$ (e.g., below the mid-depth).

Both with and without a background current, the DJL equation is formally elliptic (i.e., having no time dependence and a form dominated by a Laplacian) and strongly nonlinear due to the fact that the background density and velocity (when present) are evaluated at the upstream height (i.e., at $z - \eta$). The boundary conditions for both the sheared and constant current cases are the same: for far upstream and downstream, we assume that η tends to zero, while both the rigid lid (at $z = H$) and flat bottom must be streamlined, so that

$$\eta = 0 \text{ when } z = 0, H.$$

One interesting point to note is that in the limit of the linear stratification (or constant $N^2(z)$) the DJL linearizes. This is because the evaluation of the squared buoyancy frequency at the upstream height, $N^2(z - \eta)$, now yields a constant. The DJL with a constant $N^2(z) = N_0^2$ has been used substantially in the literature covering mountain waves in the atmosphere. Moreover, much of Long's original work in the equation involved a linearly stratified fluid. Nearly, all measurements of the coastal ocean yield density profiles that are far from linear. Similarly, much of the wave activity observed in the coastal oceans appears to involve solitary, or solitary-like waves, which are manifestly nonlinear. For these twin reasons, we will generally avoid the linearly stratified case in subsequent discussion.

Doing It Numerically Part III: DJL

The DJL equation without a background current has been solved by various methods. However, the case with a background current is, as of the time of writing, addressed only by the method developed in our group. Our method builds on the optimization procedure described in [16], which was in turn originally described in [10, 17]. The notation of this methodology is opaque, even to the general applied mathematician, and hence is presented here for the algebraically simpler case of no background current. The optimization problem is solved iteratively on the domain $0 \leq x \leq L$, $-H \leq z \leq 0$. The initial guess is obtained from KdV theory (though other guesses are possible, too) and A, and the scaled available potential energy is held fixed, while the kinetic energy is minimized. The next iteration (given the current iteration η^k and c^k) is computed as follows:

1. Solve the Poisson problem

$$\nabla^2 \eta^k = -\lambda^k S(z, \eta^k), \tag{4.13}$$

where $\lambda^k = \frac{gH}{c_k^2}$, and

$$S(z, \eta) = -\frac{\bar{\rho}'(z - \eta)\eta}{H}. \tag{4.14}$$

2. Compute λ^{k+1} by

$$\lambda^{k+1} =$$

$$\max\left[0, \frac{A - F(\eta^k) + \int_D \int S(\eta^k)\eta^k dxdz}{\int_D \int S(\eta^k)\eta^k dxdz}\right], \qquad (4.15)$$

where

$$F(\eta) = \int \int_D f(z, \eta)dxdz, \qquad (4.16)$$

and

$$f(z, \eta) = \int_0^\eta [\bar{\rho}(z - \eta) - \bar{\rho}(z - \xi)]d\xi. \qquad (4.17)$$

3. Define the new estimate of isopycnal displacement by η^{k+1} according to

$$\eta^{k+1} = \frac{\lambda^{k+1}}{\lambda^k}\eta^k, \qquad (4.18)$$

and the new estimate of the wave propagation speed according to

$$c^{k+1} = \sqrt{\frac{gH}{\lambda^{k+1}}}. \qquad (4.19)$$

This procedure is repeated until convergence criteria are met. The efficiency and accuracy of the solution procedure are thus determined by how accurately the multiple Poisson solutions in step 1 can be carried out and the integrals in step 2 can be computed. Note that neither the wave amplitude nor the wave propagation speed is specified. Instead, the kinetic energy of the disturbance is minimized under the constraint that the available potential energy is held fixed. There is nothing obvious about this procedure; however, it is readily available (codes included with this book, DJLES [17], a C implementation maintained by Kevin Lamb available by request from him) and works well in practice. A different approach, in which the DJL is considered for piecewise constant N^2 profile, is presented in [18].

Over the years, our group has developed different solvers for this problem. These have discretized the Laplacian in (4.13) using second order finite differences, as well as Chebyshev pseudospectral and Fourier spectral methods. Finite differences are likely fine for a user wishing to implement the required code from scratch. There are many standard techniques for discretizing Laplacian-type finite difference problems, and these yield extremely sparse matrices. For the Chebyshev differentiation, we have solved the resulting matrix problem employing GMRES with various preconditioning strategies, though a detailed study of the linear algebra has never been carried out. This is because the existing codes were so fast, and there was no reason to do so. A sine transform-based solver, on the other hand, allows for spectral accuracy even using moderate grids and scales well for larger grids as it does not require any explicit matrix construction. It further allows for a simple and effective refinement strategy in which the solution is iterated to convergence, and then the grid resolution is doubled. The zero-padded Fourier transform of the coarse solution is employed to initialize the fine solution, which is subsequently iterated to convergence. Finally, considerable optimization of the FFT algorithm is available, for example, as implemented in MATLAB, allowing the solver to take advantage of built-in parallelization strategies.

In the end, it is the utility of the solution that trumps the intricacies of applied mathematics behind the solution. Here it is possible to do the necessary plotting in Matlab or to use Matlab's interpolation to set up grids for use with computational fluid dynamics (CFD) codes. Indeed, the Matlab code has been used by several research groups to test CFD codes. For the purpose of these notes, we only consider test cases within Matlab. The code is split into utilities, a calculation work horse in

```
get_eta.m
```

and various drivers. The idea is that a user would only need to change the driver as they wish and perhaps either change the plotting scripts or write their own.

The switch *verbose* controls whether the algorithm reports its progress. I suggest keeping *verbose* $= 0$ unless the algorithm fails to converge. Remember failure to converge for a nonlinear equation may be the sign of bad programming, but just as often indicates a mathematical or physical property has changed. The code

```
driver_one_wave.m
```

computes a sample of a rightward propagating wave on experimental scales and makes some plots of the result. In Fig. 4.1, we show the shaded contours of the two components of the velocity field with 6 isopycnals superimposed in white for an example wave computed from the DJL equation. The isopycnals are displaced downward from their far upstream (recall upstream means the water that the wave is propagating into) height, and hence, this wave is called a "wave of depression." It is evident that the expression of the wave in the density field has a characteristic "bell" shape, like the solutions of the approximate KdV equation. The wave-induced velocity is directed upstream (into water that the wave is propagating into) above the pycnocline and downstream (or into water that the wave has already propagated through) below the pycnocline. The vertical velocity is anti-symmetric across the wave crest with downward velocities on the face of the wave and upward velocities on the downstream face. The velocity field forms a clockwise oriented vortex, with vorticity generation due to isopycnal deflection from their far upstream height (i.e., baroclinic generation). Particle (i.e., Lagrangian) tracers would move along the isopycnals or streamlines (see Box 8). Since the surface is a streamline (because of the no flux condition), a particle at the surface would only feel horizontal velocities, and since these have only one sign, it would be transported in the direction of wave propagation. A particle at the bottom would be transported in the opposite direction.

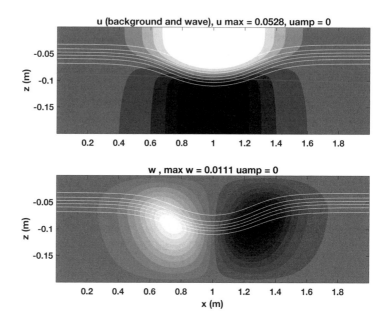

Fig. 4.1 An example of a solution to the DJL equation. The top panel shows the wave-induced horizontal velocity component (white rightward, black leftward) with six isopycnals superimposed in white. The bottom panel shows the wave-induced vertical velocity component (white upward, black downward) with six isopycnals superimposed in white

In Fig. 4.2, we show sample profiles extracted from the solution, to get some more quantitative information about the wave. The two top panels show vertical profiles at the wave crest, of horizontal velocity in the top left and density in the top right. The upstream profiles are also shown (gray); however, note in this case $U(z) = 0$. The top left panel demonstrates that the wave-induced velocity at the wave crest is essentially a shear layer across the deformed pycnocline. The magnitude of the velocity is a substantial fraction of the wave propagation speed, but $|u| < c$ at all points. The top right panel shows that at the wave crest the wave mainly moves the pycnocline downward, though a careful examination of the figure shows that the deformed pycnocline is thinner than that found far upstream.

The bottom panel shows the wave-induced horizontal velocity component scaled by the propagation speed, c, at the surface (black) and bottom (grey). The key point to note here is that while the two have a similar shape, it is not true that $u(x, 0) = \alpha u(x, -H)$ for some value of α. This means that the solution of the DJL equation is not separable, as was posited by the KdV theory.

At this point, I suggest running the script and trying out some changes. You could, for example, refine the grid and see how the execution time changed. Or if you are more physically minded, you could change the properties of the stratification profile. Of course, keep in mind that due to the nonlinearity the point is that a single wave, while nice to have, is only an example. To get an idea of trends (e.g., of

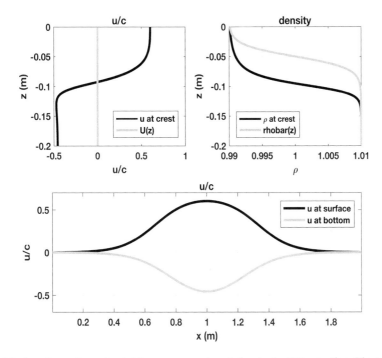

Fig. 4.2 Sample profiles extracted from an example solution to the DJL equation. The top left panel shows the wave-induced horizontal velocity component scaled by the propagation speed, c, at the wave crest (black) and $U(z)$ (grey), which is zero in this case. The top right panel shows the density at the wave crest (black) and far upstream of the wave (grey). The bottom panel shows the wave-induced horizontal velocity component scaled by the propagation speed, c, at the surface (black) and bottom (grey)

the propagation speed versus amplitude), we must compute several waves, varying some parameter in a systematic way.

As suggested by the comment above, a good start for choosing a variable to vary is to vary amplitude. This cannot be done directly in the solution algorithm. Instead, one computes waves for a range of A values, and this yields waves of different amplitudes. The script

```
driver_A_increase.m
```

gives one example of this. ISWs have an upper bound due to one of three physical scenarios, each of which expresses itself differently in solutions of the DJL equation. We begin with the same stratification as we used to compute the single wave shown in Figs. 4.1 and 4.2. In Fig. 4.3, we show the wave-induced horizontal velocity profiles as A increases. The horizontal profiles at the surface are shown in the upper panel and show that waves increase in amplitude but past a certain point broaden out without any further increase in amplitude. The width of the wave, however, continues to increase as A increases. The vertical profiles of horizontal velocity also

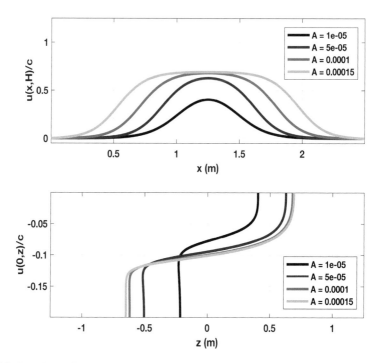

Fig. 4.3 Sample profiles as A is increased. The top panel shows the wave-induced horizontal velocity component at the surface scaled by the propagation speed, c. The bottom panel shows the wave-induced horizontal velocity component scaled by the propagation speed, c, at the wave crest. The color scheme is chosen so that larger values of A are progressively lighter

stop changing when A passes a certain value. Indeed, the two largest values of A shown have vertical profiles of horizontal velocity that are very hard to tell apart.

Figure 4.4 shows the expression of the wave in the density field as A increases. It can be seen that the waves do indeed broaden, reaching a state in which the crest region is essentially flat. This implies that the velocity field is essentially purely horizontal in this region. The limit of waves that are infinitely long is called the broadening or conjugate flow limit where the latter refers to the flow at the crest. The 1998 paper by Lamb and Wan [9] has a detailed discussion of this case, and indeed, the theory has led to an interesting set of consequences for waves over topography as explored in the next chapter and the references [23] and [25].

The broadening scenario described above is a bit surprising, if one thinks of our intuition for surface waves: larger waves generally lead to breaking. Indeed, this simplest scenario is the one predicted by the KdV equation. To see this, re-examine Boxes 5 and 7, and note that the streamfunction scales as the wave amplitude, meaning the largest wave-induced velocity scales with the amplitude, while the propagation speed scales with amplitude multiplied by the nonlinearity parameter α (typically small). Moreover, the wave width scales with the inverse of the γ parameter in Box 7, implying that larger waves are narrower.

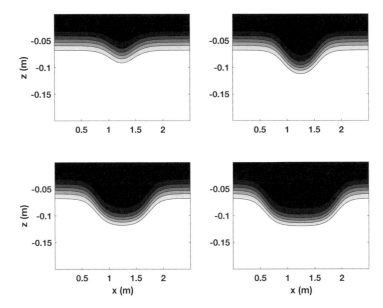

Fig. 4.4 Six shaded isopycnals for solutions of the DJL equation as A is increased. Note the broadening of the wave in the two bottom panels

In Fig. 4.5, we show the wave-induced horizontal velocity profiles as A increases for a stratification chosen so that the pycnocline center is closer to the surface, and indeed some stratification is evident all the wave to the surface. As A is increased, the profiles of horizontal velocity at the surface certainly change shape, but more importantly they surpass the wave propagation speed (recall it is solved for as part of the solution procedure). This means a particle in the wave at some time can end up ahead of the wave at a later time, and this is another way to say that breaking must occur. In fact, the DJL equation is formally not valid in such cases since the derivation assumes all streamlines connect to the undisturbed state far upstream. However, the DJL has proven useful even in this somewhat dodgy circumstance. The vertical profiles of horizontal velocity at the wave crest are shown in the lower panel. These show that the region in which $u > c$ occurs near the upper boundary, and indeed, the size of this "breaking" region can change a fair bit as A is varied.

To get an idea of the predicted wave shape when waves are breaking, the left panel of Fig. 4.6 shows the density field. Twenty isopycnals are shown in black as well as a dashed curve denoting the isopycnal found at the surface far upstream. The shape of this curve suggests there is a region of light fluid trapped inside the propagating wave. This is often referred to as a trapped core, and its realization in time dependent simulations remains a topic of active research. The DJL equation predicts that the trapped core consists of fluid lighter than that found at the surface far upstream. While not impossible, this is unlikely to be the case in nature, and a variety of prescriptions for wave properties inside the core region have been

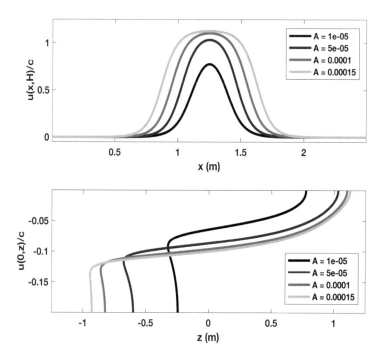

Fig. 4.5 Sample profiles as A is increased. The top panel shows the wave-induced horizontal velocity component at the surface scaled by the propagation speed, c. The bottom panel shows the wave-induced horizontal velocity component scaled by the propagation speed, c, at the wave crest. The color scheme is chosen so that larger values of A are progressively lighter. Note that $u > c$ in some cases

suggested in the literature. More discussion of the time dependent behavior of cores can be found in the last chapter of this book. If you are mathematically inclined, you might be wondering how the presence of the "core" is expressed mathematically. In the DJL, the "core" is a region with closed isopycnals. If you recall that the buoyancy frequency squared in the DJL equation is evaluated at the upstream height, you immediately conclude that waves past breaking have a region in which $z - \eta$ is evaluated outside of the physical domain. The algorithm, however, just uses the analytical continuation of the formula for the density profile.

It is quite interesting that there are two such different types of upper bound on wave amplitude. However, that is not the whole story. Since many laboratory studies of internal solitary waves use a nearly two layer stratification, there has long been speculation for what happens as the stratified layer shrinks in vertical extent. The shear in this case is more concentrated, and while a solitary wave is not one of the well-studied problems of hydrodynamic stability theory, one would at least hope that some of the theoretical results will carry over. The reader may recall that for a parallel, stratified shear flow, linear stability is guaranteed if the so-called

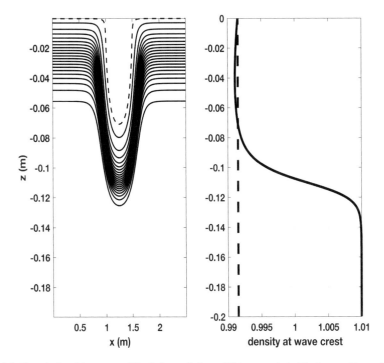

Fig. 4.6 Sample breaking waves. The left panel shows 20 isopycnals in black as well as a dashed curve denoting the isopycnal found at the surface far upstream. The right panel shows the vertical profile density at the wave crest with the value at the surface denoted by a vertical dashed line

Richardson number is above the critical value

$$Ri = \frac{N^2(z)}{U_z^2(z)} > 0.25.$$

The Richardson number, or gradient Richardson number in some references, is a ratio of the strength of the stratification to the strength of the shear [1]. $Ri < 0.25$ is not a sufficient condition for instability. However, a low Ri, especially if $Ri \approx 0.1$ or less, is a strong suggestion of instability. The idea of those internal solitary waves that are solutions of the DJL equation as the solution to an optimization problem suggests waves that may induce instability could be problematic for the solution algorithm. The script

```
driver_A_increase.m
```

allows for a choice of stratification that yields waves with a low Richardson number, and Fig. 4.7 shows sample vertical profiles of horizontal velocity and Richardson number in the upper and lower panels, respectively. It is evident that the algorithm is able to compute with low Ri. The reader can confirm by experimenting with the script that the algorithm is prone to wandering when Ri dips.

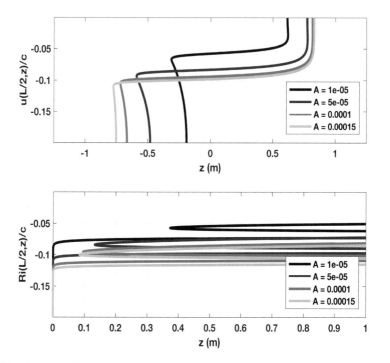

Fig. 4.7 Upper panel: sample vertical profiles of horizontal velocity at the wave crest as A increases. Lower panel: sample vertical profiles of Richardson number, Ri, at the wave crest as A increases. Note the low values for largest two A values

The bulk of the explorations presented above have been confirmed by direct numerical simulation. The 1998 paper by Lamb and Wan [9] has a detailed discussion of the flat-crested wave case, and indeed, the theory presented in this chapter has led to an interesting set of consequences for waves over topography as explored by various authors, and the following chapter.

The final tutorial script I have included is

```
driver_shear.m
```

which demonstrates the role of background shear.

Figure 4.8 shows waves with a fixed A but different linear $U(z)$ profiles. The upper row shows the density field, and the lower panel the total horizontal velocity field. A linear profile of $U(z)$ is not representative of conditions in the field but has the advantage that the shear, $U_z(z)$, is constant. Figure shows that when the shear is oriented against the direction of propagation (left column) waves that are taller and narrower, while shear oriented in the direction of propagation (right column) yields broader waves. The no shear wave is shown in the middle column for comparison.

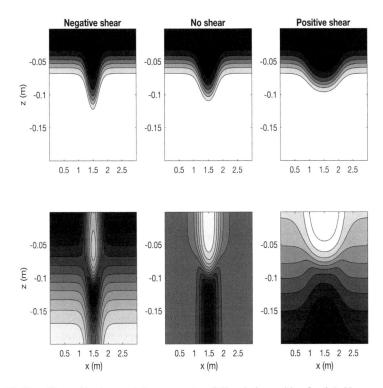

Fig. 4.8 The effects of background shear current on DJL solutions with a fixed A. Upper panel: 6 shaded isopycnals. Lower panel: 10 shaded contours of u/c saturated at ± 0.6

The density field retains its familiar solitary wave form, but the horizontal velocity contours are much trickier to interpret. Indeed, the upshot is that shear can modify the form waves take but can also do much more. It can change the type of upper bound on wave amplitude and in some cases can preclude any steady waves at all. I suggest some free play with the script will allow the reader to ascertain this for themselves. As discussed in the second mini-project, internal solitary waves with shear still have not revealed all their surprises.

Mini-Project 1: Nearly Linear Stratifications The driver

`driver_nearlin.m`

considers the question of what happens to the DJL equation and its ISW solutions when the stratification is nearly linear. When the stratification is exactly linear, the DJL equation linearizes:

$$\nabla^2 \eta + \frac{N_0^2}{c^2} \eta = 0.$$

A host of mathematical results immediately become available, and these imply that there are no ISWs in this case (feel free to dig into maximum principles at this point). But what if we have

$$\nabla^2 \eta + \frac{N_0^2 + \epsilon N^2 (z - \eta)}{c^2} \eta = 0?$$

Surely ISWs are possible, and the driver explores the wave properties of such waves. You could try pushing as close to the linear profile as you can, though you will want to be careful about the algorithm's convergence and so turn verbosity on by setting *verbose* = 1.

Mini-Project 2: Waves from Data: The second "research" driver involves actual field data. The data is from Monterrey Bay, courtesy of Ryan Walter, Cal Poly San Luis Obispo. The waves that were observed occurred in fairly shallow water, but the puzzle was not the stratification, but the extremely strong shear in the environment. The driver takes the data and interpolates to create the density and velocity fields the solver needs. However, the measurements available do not really say in which directions the waves propagate. Since the background current has both East–West and North–South components, waves propagating at different angles "feel" a different background current. The driver allows you to explore some of what happens, and indeed how in some cases, waves fail to be computed at all. In fact, the observation of waves that were not possible to reproduce with the DJL solver led to the numerical study Stastna and Walter (2014) Phys. Fluids, **26**, which explored how waves can be generated over topography when a background shear is present.

Concluding Thoughts

The perceptive reader will have noted that all the ISWs shown above were waves of depression, for which isopycnals are deflected downward from their far upstream, or rest, height. For cases with no background current, waves of elevation, for which isopycnals are deflected upward from their rest height, result when the stratification profile is reflected across the mid-depth $z = H/2$. This can be accomplished by appropriately changing the $N^2(z)$ profile (i.e., via Eq. (4.12)) in the MATLAB script.

Both from the point of view of mathematics and oceanography, the DJL equation provides a very unusual point of view on internal solitary waves. On the mathematics side, there are a couple of reasons for this. The first is a positive: the algorithm for the solution of the DJL equation works very fast and is quite robust. This allows for an exploration based approach to internal solitary waves in which "what if" questions can be answered quickly and often definitively. This is in contrast to the KdV or similar approximate theories where the specter of "how accurate is this anyway?" is always present. The second way in which the DJL equation approach is unusual is in the assumption it makes: we do not want all the solutions, and we only want traveling waves of permanent form. Since most of us learn the mathematics of linear partial differential equations first, this fundamentally nonlinear point of view is odd at first glance. On the oceanography side, most mathematical models are introduced as a means for gaining intuition for more complex simulations. As such, there is a premium on exact solutions, which are of course much easier to derive for linear models. A somewhat more subtle point is that internal solitary waves are difficult for many widely used oceanographic numerical models. This is because any model that makes the hydrostatic assumption cannot represent nonhydrostatic waves like internal solitary waves. Moreover, even if a nonhydrostatic model is used, the large geographical extent of the ocean may imply large eddy viscosities and diffusivities must be used, and these could influence, or even fundamentally alter, the waves observed in large scale models.

But you are likely wondering, if there is something concrete, we can say about how the weakly nonlinear theory like the KdV compare to the DJL equation? Well from the plots of the horizontal velocity at the surface, we have seen the velocity takes a bell shape, and indeed, one bell shape is quite difficult to tell apart from another. However, we could consider the vertical profiles of horizontal velocity at the wave crest. We saw that these effectively describe a shear layer across the deformed pycnocline. This means these are quite sensitive to both the location of the deformed pycnocline and subtleties in the vertical structure of η. In

```
driver_fits.m
```

we compute several DJL waves (staring with very small waves), and we then fit these using the linear eigenvalue problem for the vertical structure in KdV theory (Box 5). There is no unique way to perform the fit. Here the velocities at the surface are matched. Alternatively, we could minimize the root mean square error or fit the

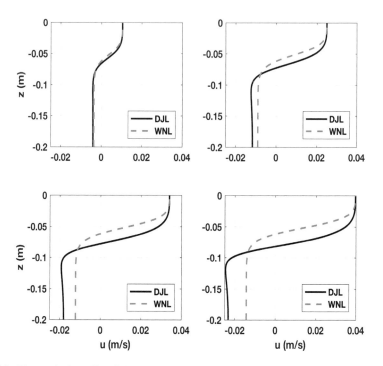

Fig. 4.9 The vertical profile of the horizontal velocity at the wave crest as given by the DJL equation and the vertical structure in the KdV theory, matching the velocity at the surface. Wave amplitude increases as the panel number increases

velocity at the bottom. In practice, the best choice is often dependent on the goals and context of the individual making the choice.

In Fig. 4.9, we show fits for waves of increasing amplitude. It is clear right away that for small waves the KdV theory does pretty well. However, very quickly as amplitude increases, the location of the shear layer is off. For the largest wave, an internal critical point is observed just below the deformed pycnocline in the DJL solution, but not in the KdV theory. The mismatch could be interpreted charitably since the KDV theory makes no guarantee of an upper bound on error. However, the issue is that physical parameters such as the Richardson number, Ri, depend on the derivative of u, meaning that small misfits in u lead to larger misfits in Ri. Thus the KdV theory is best left as a qualitative guide, with attempts at detailed fitting to observations carried out with the DJL equation.

A final point worth briefly returning to concern the conjugate flow theory, which upon re-examination, leads to a version of the DJL equation without the x dependence. To solve for a unique conjugate flow, an additional constraint must be imposed [9]. The reason the extra constraint is needed is that unlike a linear eigenvalue problem, for which the value of $\eta'(0)$ is arbitrary, the solution to the nonlinear eigenvalue problem depends on $\eta'(0)$. In practice, this amounts to solving for many pairs $(c, \eta'(0))$ that satisfy the nonlinear ODE (2.28) and boundary conditions, $\eta(0) = \eta(H) = 0$, and iterating until the auxiliary condition is satisfied to within a chosen tolerance (again see [9] for details). The interesting observation is that conjugate flow solutions may not be unique, and thus for complex stratifications (for example, those with multiple pycnoclines), there may be different families of internal solitary wave solutions. While this issue is not "tidy" enough for a mini-project, it would nevertheless make for an interesting exploration for a particularly energetic reader.

Literature 4

14. Dubreil-Jacotin ML (1932) Sur les ondes de type permanent dans les liquides heterogenes, *Atti Acad Naz Lincei, Rend Classe Sci Fis Mat Nat*, 15(6):814–919.
   The original derivation of the DJL equation.
15. Long, R,R. (1953) Some aspects of the flow of stratified fluids: A theoretical investigation, *Tellus*, 5, 42–58.
   This paper, and more than a few follow-ups, popularized the DJL equation, and especially it is linear counterpart (the N^2 constant, or linearly stratified case). In the meteorology, the DJL equation almost always appears in its linear form.
16. B. Turkington, A. Eydeland, and S. Wang (1991) A computational method for solitary internal waves in a continuously stratified fluid, *Stud. Appl. Math.* 85, 93.
   This rather difficult paper presented the optimization-based method for treating the DJL equation.
17. Dunphy, M.D., Subich, C., Stastna M. (2011) Spectral methods for internal waves: indistinguishable density profiles and double–humped solitary waves, *Nonlin. Processes Geophys.*, 18, 351–358.
   This paper presents the numerical implementation of the method in [11] that is used for all internal solitary wave calculations in this book. The latest version of the code can be found at: https://github.com/mdunphy/DJLES
18. Fructus, D., Grue. J. (2003) Fully nonlinear solitary waves in a layered stratified fluid, *J. Fluid Mech.*, 505, 323–347.
   A different point of view on the DJL equation can be realized by taking N^2 to be piecewise constant. This point of view is developed paper by the authors of this paper.

Chapter 5
Exact Internal Hydraulics

In the last chapter, we saw how making the assumption of a traveling wave of permanent form allowed us to convert the stratified Euler equations (a system of PDEs) into a single equation, namely the DJL equation. We then solved the DJL equation numerically to learn a rather large variety of things about internal solitary waves. It is worth restating that while making an a priori assumption about solution form has some mathematical precedent (e.g., Stokes' famous similarity solution for an impulsively accelerated plate under a half space of viscous fluid), it is somewhat different than the bulk of ways in which the equations of fluid mechanics have been simplified. Indeed, we have already seen an example of this more commonly applied point of view through our application of WNL, which in our context is an expansion of the stratified Euler equations in terms of two small parameters (one measuring amplitude and the other the aspect ratio). It is now fair to ask if the DJL equation provides insight into more than just propagating waves of permanent form.

In this chapter, we provide an affirmative answer to this question using the study of internal hydraulics or in other words the study of stratified flow over isolated topography. This field of study has a long history in both laboratory and mathematical contexts. It has also led to some spectacular observations in the field (e.g., Knight Inlet, British Columbia, Canada [19]). A schematic diagram is shown in Fig. 5.1. It shows a channel filled with stratified fluid that is flowing from left to right with a fixed speed U. A small hill disturbs the flow and leads to the generation of disturbances moving both upstream (right to left) and downstream (left to right). In the diagram, two lines of constant density, or isopycnals, are used to indicate that even when the hill is small in amplitude, the response may be large amplitude when the current is "just right." This is sometimes referred to as **resonant generation** of internal waves, and "just right" is taken to mean that the inflow is near one of the natural propagation speeds of waves in the system.

Mathematically, internal hydraulics is quite varied with approximate theories of all sorts (see [20] for a variety of references). Theories that neglect dispersion and discuss how signals propagate using sets of approximate, hyperbolic equations

83
M. Stastna, *Internal Waves in the Ocean*, Surveys and Tutorials in the Applied Mathematical Sciences 9, https://doi.org/10.1007/978-3-030-99210-1_5

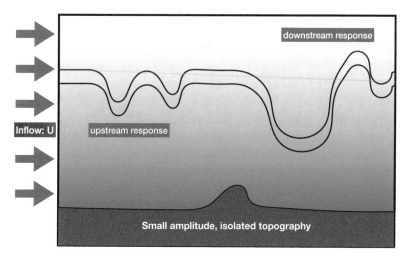

Fig. 5.1 A schematic of the flow of stratified fluid over isolated topography. Flow is from left to right, with a constant speed U. Stratification is indicated by shading and by showing two isopycnals in black. Both upstream and downstream responses are shown

have a voluminous literature. Unfortunately, they do not have a strong link with the previously discussed WNL theory since they wholly neglect all dispersive terms. There is, however, a theory of the so-called forced KdV equation, or fKdV [21], which predicts a variety of nonlinear waves both upstream and downstream of isolated topography. Moreover, when away from the topography, the fKdV theory reduces to the KdV theory discussed in previous chapters. In Fig. 5.2, we provide a schematic of the hydraulic response as a function of the inflow speed. For low inflow speeds, waves are free to propagate both upstream and downstream and are observed to be of small amplitude. When the inflow speed U is near the linear wave speed of a particular mode (mode-1 most often), a response far larger than the topography amplitude may be observed. As mentioned above, this regime is called the resonant generation regime and is labelled by the large, horizontal curly bracket in Fig. 5.2. A numerical simulation of the stratified Euler equations for the case of a series of small bumps is shown near the top of the figure. It is clear that the response in the resonant generation regime can be both complex and of a consistently large amplitude (say compared to the largest absolute amplitude of the topography). The complexity of the response is quite impressive and results from the fact that while the resonant generation mechanism is the same as what one would observe for a single isolated topography, the actual pattern observed reflects multiple waveforms, generated from multiple generating sites and subsequently interacting. At the time of writing, there is no widely available method for analyzing a field of waves like that in Fig. 5.2 and breaking it up into its nonlinear constituents. In contrast, WNL theories of internal hydraulics are tidy, and their resulting equations may be solved quickly and efficiently with modern software packages like MATLAB. However, these theories share many of the same weakness of the WNL described in Chap. 3.

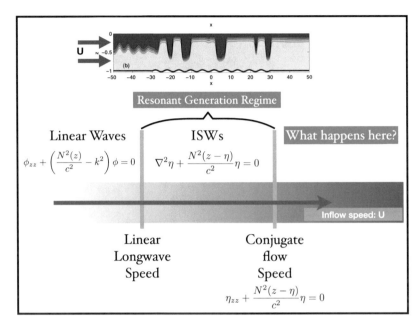

Fig. 5.2 A schematic of the dependence of internal hydraulics, on the inflow parameter U, along with the relevant equations used to describe the internal waves generated

The question that we may pose for DJL theory is then: "Is there a parameter regime for which the response over the topography can be described by an exact equation which is a variant of the DJL equation?" If the answer is affirmative, as indeed it is, then this variant of the DJL equation may be used to study such waves in an efficient manner, similarly to what was shown in Chap. 4 for internal solitary waves. Figure 5.3 shows an example of such a wave that developed in a time dependent simulation and demonstrates that the wave may be observed when the inflow speed U is close to, but larger than, the largest possible internal wave speed (or the conjugate flow speed). The conjugate flow is a limiting case of the DJL equation, for the case of flat-crested solitary waves, and is presented, along with code, in Chap. 3.

Steady DJL Theory

In Chap. 4, we have derived the DJL equation in its full form, including a shear background current for internal solitary waves. Thus, the important points for the case of exact internal hydraulics are: (i) what assumptions must be made and (ii) how is the equation and its boundary conditions modified. We consider an infinite

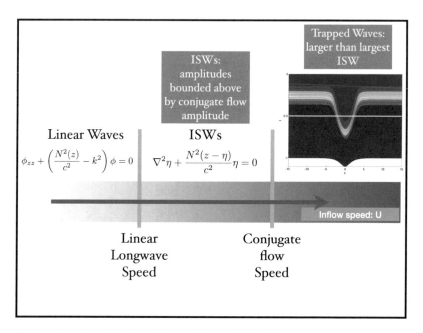

Fig. 5.3 A schematic of the dependence of internal hydraulics, on the inflow parameter U, showing the large, trapped waves typical of the truly supercritical state

strip in x, a fixed upper boundary at $z = H$ and an isolated topography given by $z = h(x)$, so that h and all its derivatives tend to zero as $|x|$ gets large. A typical example of topography is the Gaussian

$$h(x) = h_0 \exp[-(x/w_d)^2] \tag{5.1}$$

where the absolute magnitude of the topography is given by $|h_0|$. The topography may be hill-like or valley-like, depending on the sign of h_0.

We assume that far upstream the velocity field is of the form $(U, 0)$, where U is specified as part of the problem. Extensions to the case of a background shear current and that account for non-Boussinesq effects are available in the literature, but we stick to the simplest possible case herein. We also assume that the density profile far upstream is known, $\bar{\rho}(z)$, and stable, $\bar{\rho}'(z) \leq 0$ for all z. The assumption on the far upstream flow has a subtle, but important implication. It implies that the inflow U must be large enough so that any disturbance generated by the topography cannot propagate upstream. In the hydraulics literature, this is referred to as **supercritical flow**, though this is somewhat misleading since supercriticality is often taken to be determined by a Froude number based on the linear wave speed, i.e.,

$$Fr = \frac{U}{c_{linear}} > 1.$$

Since ISWs may propagate considerably faster than c_{linear} (i.e., faster than the mode-1 linear, longwave speed), we must in fact require that U is larger than any possible ISW speed. This may be determined on an *ad hoc* basis using DJL theory, but for a class of stratifications it can be computed a priori as a solution of the conjugate flow problem. We will assume that the upper bound on ISW speeds is known and labelled as c_j. We require $U > c_j$ and label such flows as **truly supercritical**. In terms of a Froude number, we would say

$$Fr^{(j)} = \frac{U}{c_j} > 1.$$

The conservation of mass, $\nabla \cdot \vec{u} = 0$, implies we can introduce a streamfunction so that $(u, w) = (\psi_z, -\psi_x)$. Since the flow is steady, and we assume $\rho = \rho_0 \bar{\rho}(z - \eta)$, Eq. (4.5) for ISWs, the density equation simply reads

$$\rho_0 \bar{\rho}'(z - \eta) J[z - \eta, \psi] = 0$$

and because $\bar{\rho}'(z)$ is not identically zero, we must have $\psi = G(z - \eta)$. Taking the limit far upstream, we find $G(z) = Uz$, and hence, $\psi = (z - \eta)U$, which is a somewhat simpler version of (4.6).

The vorticity equation written in terms of ψ and $\bar{\rho}(z)$ is considerably simpler than (4.9),

$$J[-U\nabla^2\eta, U(z - \eta)] = J[\bar{\rho}(z - \eta)g, z].$$

The right hand side may be simplified by the same, somewhat laborious, algebra as in Chap. 4, so that

$$J[\bar{\rho}(z - \eta)g, z] = J[N^2(z - \eta)\eta, z - \eta].$$

Moving all terms to one side, and using the bilinearity of J, we find

$$J[U^2\nabla^2\eta + N^2(z - \eta)\eta, z - \eta] = 0$$

so that

$$U^2\nabla^2\eta + N^2(z - \eta)\eta = F(z - \eta).$$

Taking the limit far upstream, we find $F(z) = 0$ and thus derive the DJL equation

$$\nabla^2\eta + \frac{N^2(z - \eta)}{U^2}\eta = 0. \tag{5.2}$$

As for the case of ISWs we have that η and all its derivatives vanish as $|x| \to \infty$, we also have $\eta = 0$ at $z = H$. The new boundary condition is along the bottom, namely

$$\eta(x, z = h(x)) = h(x). \tag{5.3}$$

Since U is assumed known a priori, we have a simpler problem; the DJL equation is a boundary value problem in this context. However, the domain is no longer rectangular, and this will require some discussion of appropriate numerical methods. As in the case of ISWs, a linearly stratified fluid, or one for which $N^2(z) = N_0^2$ linearizes the PDE. In fact, in the context of atmospheric internal waves, the DJL equation with $N^2(z) = N_0^2$ has a considerable literature ([22] provides an entry way into this literature), though the open upper boundary alters both the mathematics and physics of the situation to such a degree we will not discuss it further in what follows.

Steady DJL Numerics

The two key numerical questions to answer when attempting to solve the exact internal hydraulics problem are: (i) how to handle the non-rectangular geometry and (ii) how to iterate the nonlinear elliptic problem (5.2) to convergence. There are many possible methods in the literature to choose from. In this book, we consider pseudospectral methods, and for these, both a mapping method and an embedding method have been tested for point (i). Since we have shown that the embedding method is considerably more efficient [24, 25], we adopt it exclusively. Point (ii) will be handled by introducing a "time-like" variable s so that the parabolic, or heat, equation corresponding to Eq. (5.2) reads

$$\frac{\partial \eta}{\partial s} = \nabla^2 \eta + \frac{N^2(z - \eta)}{U^2} \eta. \tag{5.4}$$

If the second term on the right is interpreted as a source $R(z)$, we can apply an implicit in the time-like variable discretization

$$\left(\mathbf{I} - \frac{\Delta s}{U^2} \nabla^2 \right) \eta^{(n+1)} = R^{(n)}(z). \tag{5.5}$$

The solution will tend to the steady state, namely the solution of the DJL equation. In practice, Eq. (5.5) may be iterated until a Cauchy stopping criterion is satisfied $|\eta^{(n+1)} - \eta^{(n)}| < \epsilon$.

The embedding method, originally due to Laprise and Peltier, [22], begins by constructing the rectangular domain $[-\infty, \infty] \times [-h_D, H]$, where $h_D = min(h(x))$ (e.g., for the Gaussian example with $h_0 < 0$ $h_D = |h_0|$). An initial guess for the bottom boundary condition $\eta(z = -h_D, x) = f^{(0)}(x) = h(x)$ is chosen, and then the following iterative algorithm is applied:

1. Iterate the problem (5.5) until the stopping criterion is satisfied.
2. Define the error $e(x) = \eta(z = h(x), x) - h(x)$.
3. Define a new boundary condition $\eta(z = -h_D, x) = f^{(n)}(x) - e(x)$.

The algorithm is stopped when max $|e(x)| < \epsilon_h$ is satisfied.

Two sample codes are included with this book. The first, `DJLBCmain.m` computes a single trapped wave, while the second, `DJLBCmain_update.m`, computes multiple waves as U varies. The latter uses the previously calculated wave as an initial guess. All use a grid refinement strategy in which a solution on a coarse grid is refined at certain intervals (in practice as the outermost loop in the code). Since a pseudospectral method is employed, one can start with a fairly coarse grid.

The code displays two different diagnostics. The first is a Cauchy type error between the present η and the η from the previous iteration. This shows how quickly or slowly the algorithm is converging to a steady state. The second is the maximum absolute pointwise difference when the DJL equation is evaluated numerically using the present η. This shows how well the algorithm converges to a solution.

Steady DJL Results

We now put the numerical methods from the previous section into action. In Fig. 5.4, we show an example of the η (upper panel) and ρ (lower panel) fields for an example of a large, trapped wave over valley-like topography. It can be seen that the maximum absolute isopycnal displacement occurs at the lowest point of the topography. However, the shape of the trapped wave does not match the shape of the topography. Moreover, the solution for η is not separable in x and z, as can be seen from the tilt in the contours of η in the upper panel. The shape of the topography does, however, control the qualitative aspects of the wave, as shown in Fig. 5.5, where a double valley-like topography results in a wave with two crests.

While we know a priori that the inflow speed must preclude all upstream propagation, we do not know exactly how the trapped waves change. We expect that for very large U the DJL equation should tend to Laplace's equation for η and hence that the maximum isopycnal displacement should occur at the boundary. In Fig. 5.6, the inflow speed is varied over the range $Fr^{(j)} = 1.07 - 1.25$ (recall we are scaling by the conjugate flow speed to ensure no waves can propagate upstream). The maximum absolute isopycnal displacement for the six cases is max$(|\eta|)/H = (0.4572, 0.4287, 0.3943, 0.3511, 0.2515, 0.1100)$. The largest possible ISW (or in other words, a freely propagating wave of permanent form) for this case has an amplitude of max$(|\eta|)/H = 0.2867$. Large, trapped waves, defined as those with $|\eta| > max|\eta_{ISW}|$, can thus be observed in panels (a)–(d).

In Chap. 4, we showed that in the absence of a background shear current, waves of depression and elevation are connected by a symmetry property (4.12) when the background density profile is "flipped" across the mid-depth. For the case of internal hydraulics, while we may again "flip" the background density profile (i.e.,

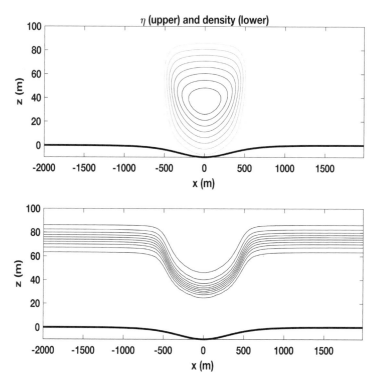

Fig. 5.4   A sample solution of a large, trapped wave over valley-like topography. η is shown in the upper panel, and the density field in the lower panel. $Fr^{(j)} = 1.0732$

in a laboratory, we could just fill the experimental tank differently), the boundary conditions have no corresponding property (i.e., it is always the bottom boundary that is deformed). Thus, an interesting question that our numerical approach can answer is, "What happens to large trapped waves when the background density profile is reflected across the mid-depth?"

Figure 5.7 shows the solution for this case with $Fr^{(j)} = 1.0732$. It is immediately evident that the trapped wave is much smaller in amplitude and indeed appears closer to a reflection of the wave in Fig. 5.6f for which $Fr^{(j)}$, and hence U, is much larger. This immediately raises the question: how do waves change as $Fr^{(j)} = U/c_j \rightarrow 1^+$ for the hill topography? Figure 5.8 shows the solution for $Fr^{(j)} = 1.0196$, a case that is very close to critical. It is evident that the trapped wave has not increased a great deal in amplitude but has increased in width. This is analogous to the case of propagating ISWs that tend to the flat crest or conjugate flow limit. For the computational scientist, it is a beautiful result for several reasons. First of all, it is emergent; I could not guess it ahead of time. Second of all, it is exact (to the extent that the stratified Euler equations apply). Third of all, it has a nice tie in with the DJL theory of ISWs.

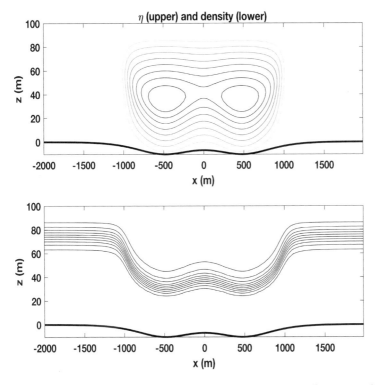

ig. 5.5 A sample solution of a large, trapped wave over a double valley-like topography. η is hown in the upper panel, and the density field in the lower panel. $Fr^{(j)} = 1.0732$

Based on this third point, here is my interpretation of why the trapped waves behave as they do: For valley-like topography, the increase in depth means that over he topography the fluid is closer to critical, and this allows for larger, trapped waves o form. For hill topography, the flow over the topography is more supercritical, and lence, waves will not get bigger but will extend upstream (and downstream) of the opography to the less supercritical region.

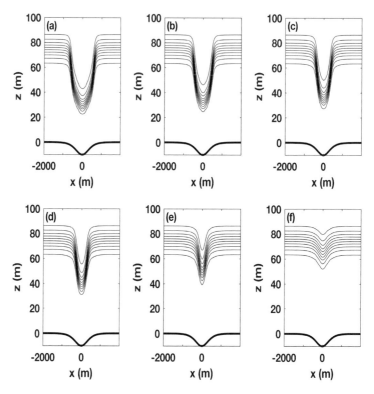

Fig. 5.6 A sample solution of a large, trapped wave over valley-like topography. The density field is shown in all panels. $Fr^{(j)} = (1.0732, 1.1090, 1.1448, 1.1806, 1.2163, 1.2521)$ in panels (**a**)–(**f**)

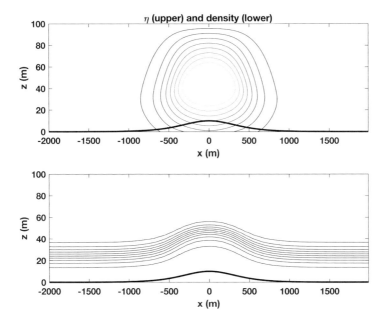

Fig. 5.7 A sample solution of a large, trapped wave over a hill topography. The background density profile is reflected across the mid-depth from that used in Fig. 5.4. η is shown in the upper panel, and the density field in the lower panel. $Fr^{(j)} = 1.0732$

Open Problems and Mini-projects

Mini-Project 1: Numerical Details The first mini-project explores some of the details of the numerical methods used in the calculation. Because spectral methods increase in accuracy with the number of grid points, one can use much lower number of grid points compared to a standard finite difference, finite element, or finite volume method. For an iterative problem, like the solution of the DJL boundary value problem, this can be exploited by first solving on a coarse grid and then interpolating onto a finer grid. In practice, this means the grid must be refined, and the previous calculation for η interpolated onto the new grid. The reader can find the details in the routine

```
make_grid.m
```

The mini-project should start with an exploration of whether refinement actually saves time or in other words solving on the fine grid from the beginning versus one or more cycles of refinement. It can then move to a direction the reader wishes. One possibility is the study of the error after refinement or in other words "can we make a better guess for the small scales."

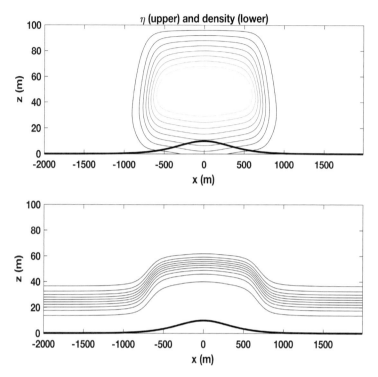

Fig. 5.8 A sample solution of a large, trapped wave over a hill topography. The background density profile is reflected across the mid-depth from that used in Fig. 5.4. η is shown in the upper panel, and the density field in the lower panel. $Fr^{(j)} = 1.0196$

Mini-Project 2: Diagnostics The second mini-project is in some ways more straightforward. The images in this chapter have focus on the isopycnal displacement η and the perturbed density field $\bar{\rho}(z-\eta)$. Using the derivation, the reader can build expressions for the streamfunction, and through it the velocities. These can be implemented using the FFT (for x) and the Chebyshev differentiation matrix (for z). The purpose of the mini-project is to build physical diagnostics, u, w, the kinetic energy, and possibly even the gradient Richardson number.

Mini-Project 3: Improved Algorithms The third mini-project is open ended. The iterative algorithm used to solve the DJL boundary value problem is quite naive. I simply created a parabolic problem out of an elliptic one and used a semi-implicit method to "time step" to steady state. I put time step in quotes since the time-like variable in this case is artificial and hence not actually time. The reader is encouraged to implement more efficient solvers for the nonlinear elliptic problem using the existing literature. When finished, these can be compared, in terms of accuracy and timing, to the codes I provide using Matlab's profiler.

Literature 5

9. Farmer, D., Armi, L. (1999) The generation and trapping of solitary waves over topography, *Science*, 5399, 188–190.
   One of many papers about Knight Inlet, British Columbia, bit at three pages, it packs quite a visual punch.
20. Stastna, M. (2011) Resonant generation of internal waves by short lengthscale topography, *Phys. Fluids*, 23, 116601.
   A good, if somewhat dates, survey of literature on resonant generation, and more broadly internal hydraulics.
21. R. Grimshaw and N. Smyth, (1986) Resonant flow of a stratified fluid over topography, *J. Fluid Mech.*, 169, 429.
   The primary reference on resonant generation in the context of internal waves, seen through the prism of KdV type theory.
22. R. Laprise and W. R. Peltier, (1989) On the structural characteristics of steady finite-amplitude mountain waves over a bell-shaped topography, *J. Atmos. Sci.*, 46, 586.
   A clever computational trick for the DJL equation over topography, and a gateway to the meteorological literature on the DJL equation with a constant N^2.
23. M. Stastna and W. R. Peltier, (2005), On the resonant generation of large-amplitude internal solitary and solitary-like waves, *J. Fluid Mech.*, 543, 267.
   This paper originally noted the possibility of large, trapped internal waves.
24. N. Soontiens, C. Subich, and M. Stastna, (2010) Numerical simulation of supercritical trapped internal waves over topography, *Phys. Fluids*, 22, 116605.
   This paper developed a generalized version of the theory discussed in this chapter and identified regimes of multiple solutions and hysteresis for realizations of the steady solutions in time dependent simulations.
25. N. Soontiens, M. Stastna, M.L. Waite (2013) Trapped internal waves over topography: Non-Boussinesq effects, symmetry ranking and downstream recovery jumps, *Phys. Fluids*, 25, 066602.
   An extension of the theory developed in this chapter to large density changes for which the Boussinesq approximation is not appropriate, as well as to inflows that are not "truly supercritical."

Chapter 6
Mode-2 Waves

The linear theory of internal waves in a channel allows for solutions of a different "mode," where mode-1 means that the structure function $\phi(z)$ is only zero at the top and bottom boundary, and each higher mode has an additional zero crossing in the interior of the domain. For example, a mode-2 wave has a single zero crossing in the interior of the channel, while a mode-4 wave would have three additional zero crossings. The weakly nonlinear theory of KdV type can be applied to higher mode waves without much trouble; all that is required is to change the propagation speed c and base the computation of the nonlinear, α, and dispersive, β coefficients on the higher mode $\phi(z)$. In contrast, the derivation of the DJL equation makes a subtle assumption about the far field structure of the isopycnal displacement, η, that may make application to higher mode solitary waves problematic. This chapter explores mode-2 waves, the example of higher mode waves that is most widely discussed in the literature. It is somewhat different in character from the other chapters because it uses time dependent simulations.

Computational fluid dynamics is a mature discipline, and good-quality software can be found from many sources. At the same time, applications of software (especially commercial software) often fall short of the rigorous standards of methods more closely tied to the analytical theory. A portion of this is natural: if we want an answer, we may be less worried about the details of how we got that answer. Indeed in complex situations, say a full simulation of internal tides in a realistic environment, concerns about how to incorporate the measurements of topography and stratification may outstrip more mathematical questions or code details. However, it should also be said that far too many times the peer review process disregards quality control of simulations completely, and the literature is full of results that are often quoted, but numerically questionable (most often due to excessive numerical viscosity/diffusion). This too is natural. Modern publishing is faster by an order of magnitude from that of past generations, and there is a natural giddiness to being able to compute answers, even when one is a bit unsure of what questions they are answers to. The code used in this chapter, called SPINS, is a

M. Stastna, *Internal Waves in the Ocean*, Surveys and Tutorials in the Applied Mathematical Sciences 9, https://doi.org/10.1007/978-3-030-99210-1_6

pseudospectral solver for the stratified, incompressible Navier–Stokes equations. It has been used in a number of contexts involving internal waves, with validation involving both theory (instability growth rates, propagation of solitary waves) and grid doubling/halving exercises [22]. This software is open source, and a link to its wiki and git repository can be found in the Literature section of this chapter. Herein we use the model in its simplest configuration: a channel with flat, free slip walls. This is an abstraction since actual tank walls are no slip, but it is closest to the assumptions behind the theory of the previous chapters.

Mode-2 Waves via Lock Release

To an experimentalist, mode-2 waves are the easiest type of internal wave to generate [23, 24], and a beautiful image courtesy of Magda Carr is shown in Fig. 6.1. A tank is stratified (for concreteness, assume with a single pycnocline), a barrier is inserted near one end, the fluid behind the barrier is mixed, and the barrier is removed. The mixed fluid intrudes into the stratification in the main tank, and a wave (possibly several waves) forms on the pycnocline. Due to the gravity current intrusion, lines of constant density are displaced upward at the top of the pycnocline and downward at the bottom of the pycnocline; a mode-2 wave is born. Experimentalists would refer to this set up as a "lock release," and a variety of improvements to the naive scenario I present have been made. For example, the stratification behind the barrier (the "lock") could be controlled, as opposed to just naively mixed. In any event, the lock release scenario is easy to implement in our numerical code. We will consider two single pycnocline stratifications, one centered at the mid-depth and the other away from the mid-depth. These are shown in Fig. 6.2. A sample lock release result is presented in Fig. 6.3 for the stratification centered at the mid-depth.

All simulations in this chapter are purely two dimensional. Three dimensional simulations of mode-2 waves have been carried out [25], and specialized techniques

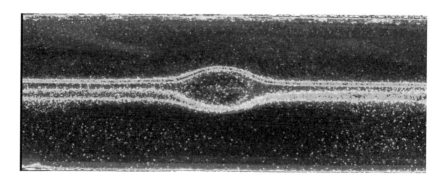

Fig. 6.1 An experimental realization of a mode-2 wave, courtesy of Magda Carr

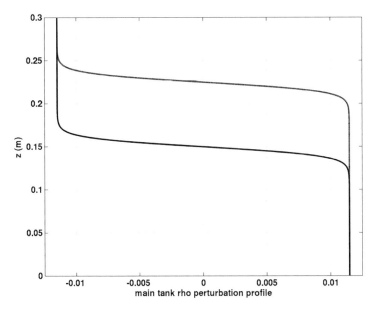

Fig. 6.2 Sample density profiles for single pycnocline stratifications. Solid line—stratification centered at the mid-depth, dashed line—stratification centered 25% of the domain below the surface

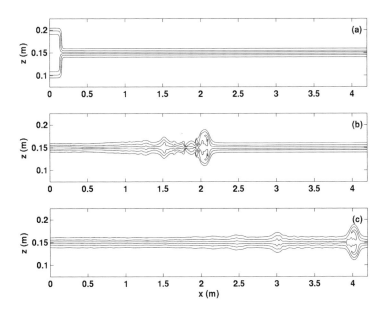

Fig. 6.3 A sample lock release simulation for the generation of a mode-2 wave by an intrusion. Pycnocline centered at the mid-depth. (**a**) $t = 0$ s, (**b**) $t = 40$ s, (**c**) $t = 80$ s

for visualization and analysis have been developed. The majority of issues for which there is a wave theoretic angle to pursue does not require three dimensional simulations, and two dimensional simulations have a far more intuitive set of visualizations. Since two dimensional simulations are computationally inexpensive, I have chosen to keep a fixed resolution of $1\,\mathrm{mm}^2$ for all simulations. Such a resolution is likely overkill for any of the physical mechanisms discussed in the remainder of this chapter. It should however be noted that most simulations do not use the physical value for the molecular diffusivity of salt. This is because it is a factor of 700 lower than the physical value of the viscosity of water [1]. We use a factor of 100 so that the pycnocline does not thicken in simulations run over a large physical time. Details of the effects of setting the diffusivty and viscosity to be equal can be found in [25]. The depth of the model tank is fixed, but the horizontal extent of the model tank varies and hence so does the number of points in the horizontal. All simulations ran in less than 1 day of computation time using 16 cores on our lab workstation, which is of the 2013 vintage. Thus what is reported is by no means cutting edge HPC. I note this because far too many papers continue to use low resolution, along with numerical methods that have considerable numerical viscosity/diffusivity. If I could implore younger readers to do one thing vis a vis their simulations, it would be "run it again with double resolution." It likely will not change your result, but it will make its reporting stronger, and on occasion may yield a surprise.

Figure 6.3 shows only a subset of the model domain, focusing on the initial state and pycnocline (panel (a)) and the evolution of the mode-2 waves. Panel (b), at 40 s, shows the typical structure of wave fissioning as the initial disturbance breaks up. There is a considerable amount of fine structure visible, including regions of overturning. Panel (c), at 80 s, shows a well separated, rank ordered mode-2 wave train. The leading wave is more than double the size of the trailing waves (two are clearly evident) and is the only region still exhibiting density overturns at this time. Shorter internal waves are evident between the first and second waves.

In order to illustrate the details of the leading wave's evolution, Fig. 6.4 shows a subdomain of the model tank that is approximately centered on the leading waves at three times. Panels (a) and (b) show the wave at the same time as Fig. 6.3, and panel (c) shows $t = 120$ s. The wave-induced horizontal velocity is shaded from blue to red, with velocity saturated at ± 0.03 m/s. Ten isopycnals are superimposed in black. It can be seen that, as predicted by linear theory, the mode-2 wave induces velocities in the direction of wave propagation in the pycnocline and against the direction of propagation above and below the pycnocline. The saturation of the velocity field is chosen so as to accentuate the gross features of the wave. This does mean that the fine scales associated with the density overturns in the center of the wave are invisible. The gross features of the density field are symmetric across the pycnocline center, and the so-called core of the wave remains active for all times shown. This in turn implies that the wave is not exactly symmetric in x across its crest, and mass continually drains from the back of the wave.

A stratification centered at the mid-depth, while convenient in the laboratory or in mathematical calculations, is unlikely to be observed in the field. Mode-2

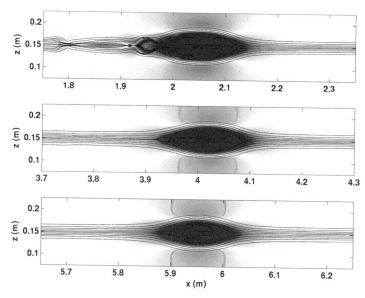

Fig. 6.4 Mode-2 waves at three stages of evolution with a subdomain chosen to be centered at the wave crest. The wave-induced horizontal velocity is shaded from blue to red, with velocity saturated at ± 0.03 m/s. Pycnocline centered at the mid-depth, and isopycnals are shown in black. (**a**) $t = 40$ s, (**b**) $t = 80$ s, (**c**) $t = 120$ s

waves have been observed in the field, [26], though they are reported far less often compared to their mode-1 counterparts. Nevertheless, it is important to ask how the generation of mode-2 waves is altered in a more representative stratification, and this has been done by both experimentalists [24] and theorists [27, 28].

Figures 6.5 and 6.6 show the counterparts of Figs. 6.3 and 6.4 for the stratification centered 25% of the domain below the surface. In Fig. 6.5, the mode-2 wave is centered to the right of the figure center in order to show the evolution of the flow behind the leading wave. Figure 6.5 clearly shows that the mode-2 waves are no longer symmetric across the pycnocline center. Interestingly, the process of mode-2 wave generation is not reduced in efficiency and indeed by 120 s yields only one mode-2 wave. This wave is followed by a wave packet, which appears to be mode-1, though this will be discussed in more detail below.

An interesting alternative point of view on the mode-2 wave's evolution is obtained by plotting the shaded enstrophy ($\omega^2/2$) in Fig. 6.7. The enstrophy is non-negative, and the monochromatic color bar clearly shows region of high vorticity. Baroclinic vorticity generation takes place in the deformed pycnocline, and vorticity with shorter length scale variations is associated with the overturns in the core region. At early times, the upper portion of the pycnocline, along with its region of baroclinic vorticity generation, extends into the wave tail. The core is active, with small scale vortex rich motions extending into the wave tail as well. With the

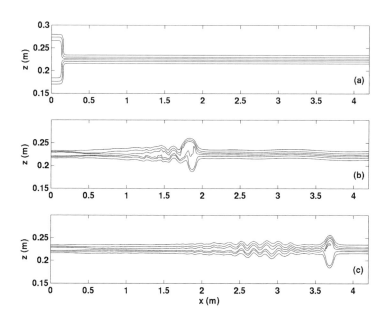

Fig. 6.5 A sample lock release simulation for the generation of a mode-2 wave by an intrusion. Pycnocline centered 25% of the domain height below the surface. (**a**) $t = 0$ s, (**b**) $t = 40$ s, (**c**) $t = 80$ s

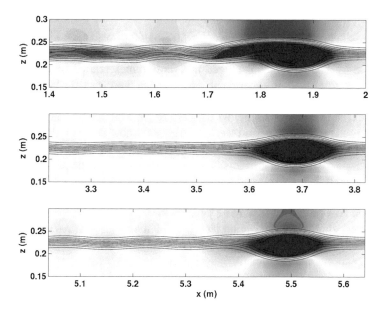

Fig. 6.6 Mode-2 waves at three stages of evolution with a subdomain chosen to be centered at the wave crest. The wave-induced horizontal velocity is shaded from blue to red, with velocity saturated at ± 0.03 m/s. Pycnocline centered 25% of the domain height below the surface, and isopycnals are shown in black. (**a**) $t = 40$ s, (**b**) $t = 80$ s, (**c**) $t = 120$ s

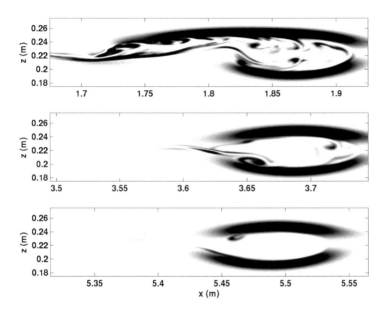

Fig. 6.7 Enstrophy in the core region and the region immediately behind the wave. Pycnocline centered 25% of the domain height below the surface. (**a**) $t = 40$ s, (**b**) $t = 80$ s, (**c**) $t = 120$ s

passage of time, the core becomes nearly quiescent, with activity only near the core boundaries. The wave also decreases in width.

The above discussion, while qualitative in nature, lays the ground work for understanding the generation problem of multiple wave modes. If we recall the linear eigenvalue problem, reproduced here for the reader's convenience,

$$\phi_{zz} + \left(\frac{N^2(z)}{c^2} - k^2 \right) \phi = 0$$

with $\phi(0) = \phi(H) = 0$, we note that in the longwave limit, the standard Sturm–Liouville theory tells us that $c_{lw}^{(1)} > c_{lw}^{(2)} > c_{lw}^{(3)} > \ldots$ and that $c^{(n)}(k)$ is a decreasing function of k for each mode n. This means that a long mode-2 wave can have the same phase speed as a shorter mode-1 wave, or more to the point, a large amplitude mode-2 internal solitary-like wave (we are hedging our bets a bit with the modifier "solitary-like") can propagate at a speed that is the same as shorter mode-1 waves.

In a numerical simulation, we have access to the velocity and density fields, as opposed to the solutions of the eigenvalue problem; thus the question becomes "how do we determine which mode contributes to a particular disturbance?" This can be a complicated problem if we wish to consider all modes, but if we only wish to sort between mode-1 and mode-2, we can appeal to the modal structure: Mode-2 has an additional zero crossing in its vertical structure. Since the wave-induced horizontal

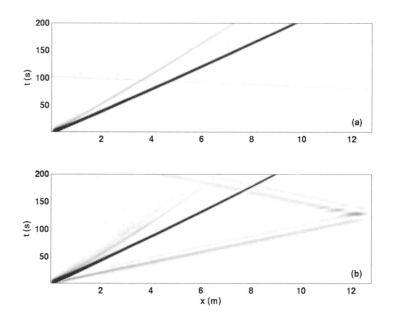

Fig. 6.8 Hovmöller plot of the mode index MI. (**a**) Pycnocline centered at the mid-depth, (**b**) Pycnocline centered 25% of the domain height below the surface

velocity $u = \psi_z \propto \phi'(z)$, this means that the product of u at the surface and bottom will be positive for mode-2 waves and negative for mode-1 waves. We thus define

$$MI(x, t) = \frac{u(x, z = H, t)u(x, z = 0, t)}{U^2} \qquad (6.1)$$

where U is a scaling factor (e.g., the mode-1 long wave speed). $MI(x, t)$ can be shaded in what is called a space–time, or Hovmöller, plot.

Figure 6.8 shows the Hovmöller plot of the mode index MI for the two stratifications. The entire model tank length is shown. Panel (a) shows the case of the pycnocline centered at the mid-depth. It is clear that only mode-2 waves form, with three evident at this saturation level. Panel (b) shows the case of the pycnocline centered 25% of the domain height below the surface and is clearly profoundly different. A single mode-2 wave dominates the evolution, consistent with Fig. 6.6. Surprisingly, the second most important feature is a pair of mode-1 waves that propagate much faster than the mode-2 wave (essentially at $c_{lw}^{(1)}$). The mode-1 long waves reflect off the right boundary and propagate leftward, passing through the mode-2 wave with both wave types largely unchanged. The mode-2 wave is trailed by an expanding fan of waves that are difficult to make out at the saturation level chosen. Nevertheless, the mode diagnostic is remarkably useful. It shows clearly that the mode-2 waves generated are coherent over the entire simulation, consistent with their "solitary-like" moniker. Moreover, because we have examined only single pycnocline stratifications, the mode-2 waves contained a "core" region with density overturns, and hence, they could not be truly "solitary."

There remain two outstanding challenges. The first relates to the mode diagnostic $MI(x, t)$, and its ability to paint an accurate picture of the activity at the pycnocline, and the second relates to the question of whether overturning is in some sense an essential aspect of mode-2 dynamics.

Mode Diagnostics and Mode-1 Tails

In Fig. 6.9, we revisit the mode-2 wave generated via lock release at three late times. The figure shows the shaded horizontal velocity field saturated at ± 0.01 m/s. Six isopycnals are overlaid over the horizontal velocity field in green. The subdomain center has been shifted horizontally so that the leading wave sits at the right side of the subdomain shown, and care has been taken that the long mode-1 waves that ran out in front of the leading mode-2 wave have not had time to reflect and return into the field of view. It is clear that the mode-2 wave is trailed by a mode-1 tail. A trailing, smaller mode-2 wave can be seen exiting the subdomain on the left in the top and middle panels. The mode-1 nature of the tail can be established from the "vertically in phase" nature of the isopycnals' displacement from their rest height (isopycnals shown in green). It can be seen that while the velocity field of the leading wave reaches well outside of the vertical range of the subdomain selected, the same cannot be said for the horizontal velocities induced by the mode-1 tail. Since the vertical extent of the subdomain is considerably smaller than that of the full domain, one gets the sense that the mode diagnostic measure $MI(x, t)$, while ideal for long waves and solitary waves of both the first and second modes, would not be an ideal tool for detecting the mode-1 tail. Of course, we could simply take horizontal slices at more appropriate heights chosen a posteriori, say $z = 0.18$ and 0.26 m in this case. But at the moment it is unclear if an a priori diagnostic measure is available.

In Fig. 6.10, we show the performance of the mode index $MI2(x, t)$ that is defined at a posteriori chosen values of z. It turned out that the above guesses of $z = 0.18$ and 0.26 m needed a little fine tuning, and the index was built on profiles at $z = 0.188$ and 0.25 m. The scaling factor was also chosen by trial and error, with a value of 0.005 m/s used in the figure. It can be seen that this mode index yields very large values of the leading mode-2 wave and much smaller, but consistently negative (meaning mode-1) values for the waves in the tail. In panel (a), the trailing mode-2 wave can be seen at the left end point of the subdomain.

With the mode-1 nature of the tail established, we can move to identify its wavelength. This can be done by a variety of means, and direct inspection is likely the easiest, yielding a wavelength of roughly 0.1578 m. Similarly, since the leading mode-2 wave induced by far the largest magnitude surface currents, a leading mode-2 wave speed can be estimated as approximately 0.0438 m/s. Linear theory predicts that mode-1 waves with a wavelength of 0.1578 m propagate with a phase speed of 0.042 m/s, which is very good agreement. Linear theory can also be used to estimate

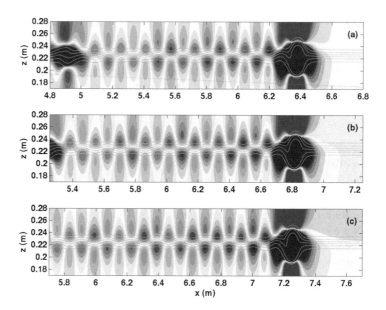

Fig. 6.9 Mode-2 waves with trailing mode-1 tails. The wave-induced horizontal velocity is shaded from blue to red, with velocity saturated at ±0.01 m/s. Pycnocline centered 25% of the domain height below the surface, and isopycnals are shown in green. (a) $t = 140$ s, (b) $t = 150$ s, (c) $t = 160$ s

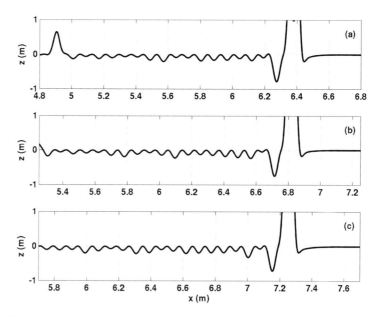

Fig. 6.10 Mode index $MI2(x, t)$ for three different times. The x values for the u profiles are $z = 0.188, 0.25$ m, and the scaling factor is 0.005 m/s. (a) $t = 140$ s, (b) $t = 150$ s, (c) $t = 160$ s

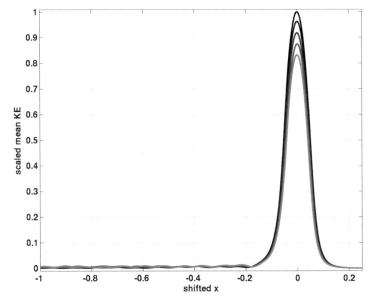

Fig. 6.11 Vertically averaged KE scaled by the maximum at $t = 120$ s. The horizontal axis is shifted so that the maximum occurs at $x = 0$. The time interval $t = 120$ to 160 s is shown in 10 s intervals. $t = 120$ s is shown in black with lighter shades of gray for each later time

the group speed associated with these linear waves, and this yields an estimate of 0.015 m/s. This means that the tail of mode-1 waves remains locked to the leading mode-2 waves but at the same time continually drains energy from the leading wave (i.e., the energy in the mode-1 tail cannot keep up with the mode-2 wave). Thus the leading mode-2 wave is not a true solitary wave but a solitary-like wave (see [28] for a more complete perturbation theory). The precise rate of decay of the solitary-like wave can be quantified in a number of ways. In Fig. 6.11, we do so via the vertically averaged kinetic energy that is scaled by the maximum value at $t = 120$ s, the earliest time shown in the figure. It can readily be seen that the leading wave decays in terms of its kinetic energy. Basic curve fitting indicates that a linear decay is a good model for the decay, with a residual on the order of 10^{-6} with a cubic fit yielding only a slight improvement (residual on the order of 10^{-7}).

The process of mode-2 generation via lock release is very closely related to that of mode-1 waves, but mode-1 solitary waves can outrun all other waves, tending to a true solitary wave, one that is described by the DJL equation. Mode-2 waves are thus only "solitary-like." One outstanding question is whether the wave tail can ever lead to a modified shape of the leading waves. This will be the subject of the next subsection.

Exotic Mode-2 Waves

The results of the last section established how the dynamics of mode-2 internal solitary-like waves differs from the dynamics of mode-1 internal solitary waves However, the mode-2 waves we have seen to date were more or less symmetric across the pycnocline center, and the slow energy drain by the mode-1 tail was manifested by a gradual decay in amplitude, as opposed to a violent collapse of the mode-2 waves. Nevertheless, one suspects that the manner in which the DJL equation is recast as a variational problem for mode-1 waves will not have an analogous argument for mode-2 waves, and this begs the question of whether even more exotic possibilities lie in store for mode-2 waves.

The study of nonlinear waves includes phenomena that are far more exotic than the simple, bell-shaped soliton solutions of the KdV equation. However, the key reason why internal solitary waves are of practical interest is precisely the fact that they are very easy to generate in the lab (via lock release) or in the ocean (via tidal flow over topography). Thus, the task we wish to set out for ourselves is to dynamically generate mode-2 waves that are qualitatively different than those of the previous section. We begin by considering two pycnoclines. This allows for two different (though coupled) wave guides and has the added attraction of precluding the formation of a recirculating core. If we expect the symmetry breaking phenomenon to be a nonlinear one, the choice of an initially large amplitude makes sense as well. The rest is down to numerical exploration.

In Fig. 6.12, we show the results of that exploration. Two pycnoclines centered at 20% and 30% of the domain height below the surface are perturbed by a large amplitude, wide disturbance, and allowed to evolve. While no overturning is observed, and aspects of the evolution are familiar (long mode-1 waves, followed by a main mode-2 solitary-like wave and a tail of short waves), the structure of the leading mode-2 wave is clearly different from that of the previous section. Only the upper half and leftmost 80% portion of the computational domain is shown.

In Fig. 6.13, we focus on the leading mode-2 wave. The times shown are chosen so that the leading, long mode-1 waves have not had time to reflect off the far wall and return to interfere with the mode-2 wave. The subdomain shown has been further restricted, with a horizontal extent of 4 meters and a vertical extent of 15 centimeters. The viewing window has been manually adjusted to have the leading mode-2 wave more or less centered at two thirds of the horizontal extent. In this way, the leading wave and the tail are shown. It is clear that at 60 s the leading wave has not fully adjusted, and the tail is not fully formed. By 120 s, the leading mode-2 wave has an upper pycnocline deformation that takes the form of a flat-crested internal solitary wave and a lower pycnocline deformation that consists of three shorter waves. The tail has begun to form in earnest, though the eye is drawn to irregularities in its structure. By 180 s, the leading mode-2 wave has decreased in size and width. The upper pycnocline deformation, while still broad crested, has lost its flat-crested central region. The lower pycnocline deformation now consists

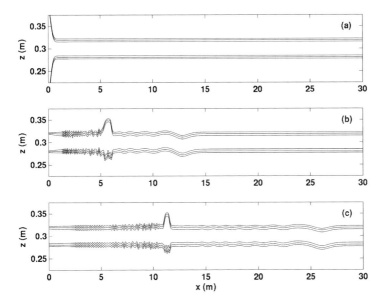

Fig. 6.12 The density field (six isopycnals in black) for a sample lock release simulation for the generation of an "exotic" mode-2 wave. The two pycnoclines are centered at 20% and 30% of the domain height below the surface. (**a**) $t = 0$ s, (**b**) $t = 150$ s, (**c**) $t = 300$ s

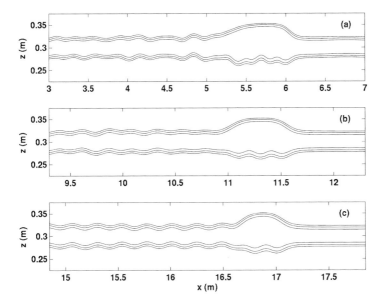

Fig. 6.13 The density field (six isopycnals in black) for a sample lock release simulation for the generation of an "exotic" mode-2 wave. The two pycnoclines are centered at 20% and 30% of the domain height below the surface. (**a**) $t = 150$ s, (**b**) $t = 300$ s, (**c**) $t = 450$ s

of two shorter wave crests, and these are seen to form the front portion of the tail
The tail takes on the form of a sinusoid with a single wavelength.

The structure of the leading mode-2 wave is revisited in Fig. 6.14 in which
we focus on the horizontal velocity field. A clear asymmetry between the region
above the two deformed pycnoclines and below the two deformed pycnoclines is
evident. The wave-induced velocities are much stronger above the two deformed
pycnoclines. This profound asymmetry appears odd at first but is in fact consistent
with linear theory, the results of which are shown in Fig. 6.15. Panel (a) shows the
vertical structure functions and $N^2(z)$ profile, while panel (b) shows the horizontal
component of the wave-induced velocity. It can be seen that for the mode-1 waves,
both the strong near surface wave-induced currents in the direction of propagation
and the slightly weaker near bottom wave-induced currents against the direction of
propagation remain unchanged by the double pycnocline stratification, though there
are two high shear regions, one for each pycnocline. The mode-2 wave induces
a strong flow in the direction of wave propagation in the region between the two
pycnoclines and a weaker flow against the direction of wave propagation above and
below the two pycnoclines. In the case of the region below the lower pycnocline,
this flow is very weak (though the relative change from the flow between the two
pycnoclines is still sizable), consistent with the numerical experiments shown in
Fig. 6.14. Linear theory predicts the mode-2 propagation speed to be roughly 35%
of the mode-1 propagation speed. However, the linear longwave mode-2 speed
underestimates the propagation speed of the exotic mode 2 waves by around 30%.

One could ask whether linear theory provides an explanation for the fact that the
bottom pycnocline has a larger amount of short wave activities, especially in the
body of the mode-2 wave. A quick calculation shows that short waves computed
for the two pycnocline stratification are in fact symmetric for the two pycnoclines.
This is intuitively sensible, since shorter waves are more trapped near the two
pycnoclines, and thus are less influenced by the fact that the two pycnoclines are
well away from the mid-depth. However, there is a further possibility by which the
mode-2 wave can affect linear theory and that is through the shear in the wave-
induced mode-2 velocity profile. Using the velocity profile shown in panel (b) of
Fig. 6.15 as the background flow in the Taylor–Goldstein equation indeed shows
that the primary effect of the background current is to shift the focus of the mode-1,
short wave structure to the lower pycnocline; see Fig. 6.16. Thus the exotic mode-2
wave can be interpreted as the mode-2 solitary-like wave with a mode-1 tail of the
previous section, with the tail penetrating into the mode-2 wave body, where it is
modulated by the mode-2 wave currents so that it is manifested preferentially on the
lower pycnocline.

The above argument could be described as "linear thinking." To test the linearity
of this scenario, we must change the amplitude of the mode 2 wave. However, since
the wave develops naturally from the initial conditions, this is not a simple matter
of tuning a parameter. Figure 6.17 compares two different amplitudes of the initial
disturbance. The large amplitude case in panel (a) is that discussed above, with panel
(b) showing a case with the initial amplitude reduced by one third. It can be seen that
a much narrower mode-2 wave is observed in panel (b), and while the behavior of the

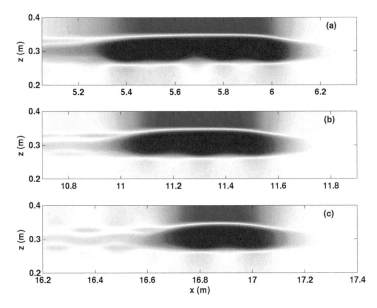

Fig. 6.14 The shaded horizontal velocity field saturated at ±0.02 m/s for a sample lock release simulation for the generation of an "exotic" mode-2 wave. The two pycnoclines are centered at 20% and 30% of the domain height below the surface. (a) $t = 150$ s, (b) $t = 300$ s, (c) $t = 450$ s

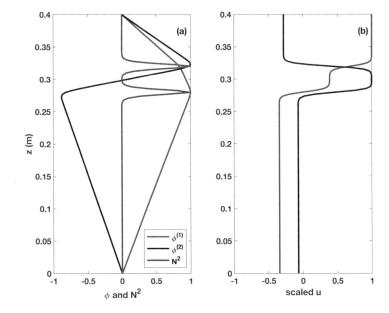

Fig. 6.15 Linear theory for the undisturbed stratification used in the exotic mode-2 numerical experiment. (a) Scaled N^2 in red, mode-1 in blue, and mode-2 in black, (b) scaled horizontal component of the wave-induced velocity for mode-1 (blue) and mode-2 (black)

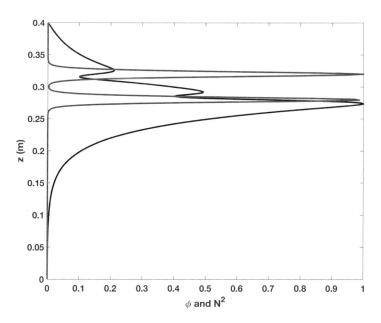

Fig. 6.16 Linear theory for the stratification used in the exotic mode-2 numerical experiment and a background current based on the mode-2 current, wavelength 0.2 m. Scaled N^2 in red, mode-1 structure in black

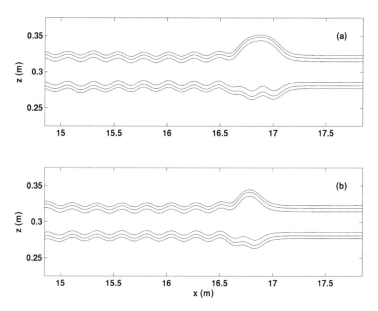

Fig. 6.17 The density field (six isopycnals in black) for two sample lock release simulations for the generation of an "exotic" mode-2 wave by an intrusion. The two pycnoclines are centered at 20% and 30% of the domain height below the surface. (**a**) Large amplitude $t = 450$ s, (**b**) amplitude reduced by one third, $t = 450$ s

ail is more or less consistent with the mechanism described above, the length scale of the waves in the mode-1 tail is such that the waves along the bottom pycnocline no longer fit neatly into the body of the leading mode-2 wave. Thus the primary way in which nonlinearity is expressed in this scenario is through the broadening of the mode-2 wave as the amplitude of the initial disturbance is increased. Indeed, evidence for this point of view can be seen, in retrospect, in Fig. 6.13 in the sense that while panel (b) has three clear crests along the bottom pycnocline in the leading mode-2 wave, panel (c) in which the mode-2 wave has decreased in length has only two. A detailed comparison of the line of thought in this chapter, and more generally simulations of nonlinear mode-2 waves using the full stratified Navier–Stokes equations, with MCC theory of the type in [11] remains to be carried out. The reader interested in novel perturbation methods should investigate [28], in which a weakly nonlinear theory for the mode-2 wave and its mode-1 tail is developed, using asymptotics beyond all orders.

Literature 6

22. Subich, C. J., Lamb, K. G., and Stastna, M.:(2013) Simulation of the Navier-Stokes equations in three dimensions with a spectral collocation method, *Int. J. Numer. Meth. Fl.*, 73, 103–129, https://doi.org/10.1002/fld.3788.
   The reference for the numerical model SPINS used in this chapter. The model wiki, including instructions on how to get a copy from git, can be found at https://wiki.math.uwaterloo.ca/fluidswiki/index.php?title=SPINS

23. T. Maxworthy (1980) On the formation of nonlinear internal waves from the gravitational collapse of mixed regions in two and three dimensions, *J. Fluid Mech.* 96, 47–64.
   A classical reference on the experimental realizations of mode-2 internal solitary waves.

24. Carr, M., Davies, P. A., and Hoebers, R. P. (2015) Experiments on the structure and stability of mode-2 internal solitary-like waves propagating on an offset pycnocline, *Phys. Fluids*, 27, 046602.
   A more current, experimental realization of mode-2 waves, and a demonstration of the effects of moving the pycnocline center away from the mid-depth.

25. Deepwell, D. and Stastna, M. (2016) Mass transport by mode-2 internal solitary-like waves, *Phys. Fluids*, 28, 056606.
   Numerical simulations of transport by mode-2 waves, and a rather broad discussion of the literature on mode-2 internal solitary waves. The issue of how to deal with the extremely low molecular diffusivity of salt is also discussed.

26. Shroyer, E. L., Moum, J. N., and Nash, J. D. (2010) Mode 2 waves on the continental shelf: Ephemeral components of the non- linear internal wavefield, *J. Geophys. Res.*, 115, C07001.
   The most widely quoted set of observations of mode-2 waves in the ocean.

27. Olsthoorn, J., Baglaenko, A., and Stastna, M.(2013) Analysis of asymmetries in propagating mode-2 waves, *Nonlin. Processes Geophys.*, 20, 59–69.
   Theoretical/numerical discussion of the effects of moving the pycnocline away from the mid-depth.

28. Akylas, T.R., Grimshaw, R.H.J. (1992) Solitary internal waves with oscillatory tails, *J. Fluid Mech.*, 242, 279.
The weakly nonlinear theory description of the mode-2 wave with a mode-1 tail using asymptotics beyond all orders.

Chapter 7
Concluding Thoughts

The last chapter's foray into time dependent simulations seems like a good place to finish our tour of the theory of nonlinear internal waves. The tools available to describe nonlinear internal waves have certainly become far more widely available over the past decade or so, and the frontier for exploration tilts toward the application of theory. Nevertheless, clever initializations, especially those rooted in theory, remain relatively rare in the literature. In some sense, it is the fact that they spontaneously appear almost from any initialization that makes mode-1 internal solitary waves such an interesting phenomenon to study. Thus authors can hardly be faulted for simply perturbing a pycnocline and allowing waves to evolve. However, numerical simulations can be far more efficient if even a little bit of theory is applied; for example by using a DJL solver to set the length scale of the initial perturbation.

While any newcomer to nonlinear internal waves can certainly learn plenty from the previous Chapters, there is much more that is available in the literature. There have been three reviews in the Annual Review of Fluid Mechanics in the past 15 years [29–31], which can provide a springboard into the literature (I would say "excellent reviews" but as I am a co-author on the most recent of the three, this seemed a bit self-serving). Each focuses on different aspects of the generation, propagation, and dissipation of internal waves. The recent Encyclopedia entry [32] provides a perspective on ISWs and turbulence, a huge topic that will only grow in importance as large 3D simulations become the norm.

As useful as they are, reviews are in some sense agnostic, since their job is to quote the broadest swath of literature; in comparison I have tried to take a stand on various issues as I explained them in the previous chapters. I hope the reader has found some of that useful, and I wish to briefly note that the point of taking a stand is not always to ensure the reader agrees. In many ways science works best when there is a disagreement that allows for truly novel progress.

The topic of waves with trapped cores made several appearances in this book and is an example of growth through different perspectives. Pioneering theoretical work suggested cores should be quiescent [33], while early numerical solutions

M. Stastna, *Internal Waves in the Ocean*, Surveys and Tutorials in the Applied Mathematical Sciences 9, https://doi.org/10.1007/978-3-030-99210-1_7

suggested they were not [34]. Experiments, performed at much lower Reynolds numbers compared to those that occur in the field, suggested that in the experimental setting cores were more or less quiescent [35, 36] perhaps with some activity at the edges of the core. Recently, motivated by field measurements in the South China Sea, a new core type was investigated theoretically [37] and using 3D numerics [38], providing a novel direction to what is a well established topic. As an aside, the work in [35] considers cores for mode-1 waves, but many of the ideas therein translate to cores in mode-2 waves discussed above in the chapter on mode-2 waves.

Towards this goal, in the remainder of the text I wish to outline one area that I believe provides fertile ground for exploration that would advance theory. I want to begin with the well known dispersion relation of Poincaré waves in shallow water theory on the f-plane [1]

$$\sigma = \sqrt{f^2 + gHk^2}. \tag{7.1}$$

Here σ is the frequency, k the wave number, H the layer depth, f the rotation parameter, and g is the acceleration due to gravity. Recall, the so-called shallow water speed is given by $c_{sw} = \sqrt{gH}$. These waves are dispersive, since

$$c_p = \frac{\sigma}{k} = \sqrt{\frac{f^2}{k^2} + gH} \tag{7.2}$$

and

$$c_g = \frac{d\sigma}{dk} = \frac{gHk}{\sqrt{f^2 + gHk^2}} \neq c_p. \tag{7.3}$$

This is an instructive example of the non-physical nature of the phase speed, since as $k \to 0^+$ we have $c_p \to \infty$, while $c_g \to 0$. The infinite phase speed is not, as a point of fact, some violation of Einstein's Theory of Relativity, because the group speed is responsible for energy transport. With only a modest increase in algebra a similar result can be derived for linear internal waves on the f-plane. The fact that c_p achieves all values between c_{sw} and ∞ means that rotation, even weak rotation typical of the Earth, has the potential to greatly affect wave dynamics.

Consider, one final thought experiment: A mode-1 internal solitary wave is generated by lock release in a very long, rotating tank. In the non-rotating theory this wave propagates faster than any other wave and hence "runs away" from the messy state associated with the lock release. However, with rotation long linear waves may have a phase speed that matches the propagation speed of the putative internal solitary wave, leading to a loss of energy, and of true "solitary" status. This is directly analogous to the discussion of mode-2 solitary-like waves in the last Chapter.

In terms of the weakly nonlinear theory, rotation leads to the so-called Ostrovsky equation (see [29], and especially [39]), which is a KdV type equation, though with a number of pathologies. Of course this is perfectly understandable given the fact that the phase speed grows without bound, and an interesting challenge for budding theoreticians would be to find a rotating version of the BBM (which regularizes short wave, linear dispersion) that regularizes both short and long wavelengths. Rotation has profound effects on the trapped waves over topography, and while I am not aware of observations corroborating the numerical experiments, truly supercritical flow modified by rotation can yield highly nonlinear, steady waves tens of kilometers from the generation site [40]. In a laboratory setting, the presence of sidewalls implies that other wave types, possible only in the rotating frame, like Kelvin waves may result [41]. It is at this point that I leave the explorations in the capable hands of the reader, and I look forward to reading about the results in the literature.

Literature 7

29. Helfrich KR, Melville WK. (2006) Long nonlinear internal waves, *Annu. Rev. Fluid Mech.*, 38, 395–425.
30. Lamb KG. (2014) Internal wave breaking and dissipation mechanisms on the continental slope/shelf. *Annu. Rev. Fluid Mech.*, 46, 231–54
31. Boegman, L., Stastna, M. (2019) Sediment resuspension and transport by internal solitary waves. *Annu. Rev. Fluid Mech.*, 51, 129–154.
32. Lamb, K. G., Lien, R.-C. and Diamessis, P. J. (2019) Internal Solitary Waves and Mixing. In *Encyclopedia of Ocean Sciences. 3rd edition*, **Vol. 3**, 533–541, Elsevier.
33. Derzho, O. G. and Grimshaw, R. (1997) Solitary waves with a vortex core in a shallow layer of stratified fluid, Phys. Fluids, 9, 3378–3385, https://doi.org/10.1063/1.869450.
34. Lamb, K. G.(2002) A numerical investigation of solitary internal waves with trapped cores formed via shoaling, J. Fluid Mech., 451, 109–144, https://doi.org/10.1017/S002211200100636X.
35. Carr, M. King, S.E. and Dritschel, D.G. (2012) Instability in internal solitary waves with trapped cores, Phys. Fluids, **24**, 016601, https://doi.org/10.1063/1.3673612.
36. Luzzatto-Fegiz, P. and Helfrich, K. R. (2014) Laboratory experiments and simulations for solitary internal waves with trapped cores, J. Fluid Mech., 757, 354–380, https://doi.org/10.1017/jfm.2014.501.
37. He, Y., Lamb, K. G., and Lien, R. C. (2019) Internal solitary waves with subsurface cores, J. Fluid Mech., 873, 1–17, https://doi.org/10.1017/jfm.2019.407.
38. Rivera-Rosario, G. Diamessis, P.J. Lien, R-C/ Lamb, K.G. and Thomsen, G.N. (2020) Formation of Recirculating Cores in Convectively Breaking Internal Solitary Waves of Depression Shoaling over Gentle Slopes in the South China Sea, J. Phys. Oceanogr. **50**, 1137–1157.
39. Grimshaw, R., Ostrovsky, L., Shrira, V.I. and Stepanyants, Y.A. (1998) Long nonlinear surface and internal gravity waves in a rotating ocean, *Surv. Geophys.*, 19, 289.
40. Stastna, M., Subich, C. and Soontiens, N. (2012) Trapped disturbances and finite amplitude downstream wavetrains on the f–plane, *Phys. Fluids*, 24, 106601.
41. Deepwell, D., Stastna, M., Subich, C. (2018) Multi-scale phenomena of rotation–modified mode-2 internal waves, *Nonlin. Proc. Geophys.*, 25, 217–231.

Index

||| || || ||||||| | ||| || |||| ||| |||||| ||| |||| ||

Printed in the United States
by Baker & Taylor Publisher Services